S0-CKI-789

Proceedings of the Institute of General Physics
Academy of the Sciences of the USSR
Series Editor: A.M. Prokhorov

Proceedings of the Institute of General Physics
Academy of the Sciences of the USSR
Series Editor: A.M. Prokhorov

Volume 1 **Oceanic Remote Sensing,** Edited by F.V. Bunkin and K.I. Volyak.

Volume 2 **Laser Raman Spectroscopy in Crystals and Gases,** Edited by P.P. Pashinin.

Volume 3 **The Magnetic and Electron Structures of Transition Metals and Alloys,** Edited by V.G. Veselago and L.I. Vinokurova.

Volume 4 **Laser Methods of Defect Investigations in Semiconductors and Dielectrics,** Edited by A.A. Manenkov.

Volume 5 **Fiber Optics,** Edited by E.M. Dianov.

Volume 6 **The Nonlinear Optics and Acoustics of Fluids,** Edited by F.V. Bunkin.

Volume 7 **Formation and Control of Optical Wavefronts,** Edited by P.P. Pashinin.

Volume 8 **Problems of Lithography in Microelectronics,** Edited by T.M. Makhviladze.

Volume 9 **Selective Laser Spectroscopy of Activated Crystals and Glasses,** Edited by V.V. Osiko.

Proceedings of the Institute of General Physics
Academy of the Sciences of the USSR
Series Editor: A.M. Prokhorov
Volume 4

LASER TECHNIQUES FOR INVESTIGATION OF DEFECTS IN SEMICONDUCTORS AND DIELECTRICS

Edited by A.A. Manenkov

Translated by Kevin S. Hendzel

NOVA SCIENCE PUBLISHERS
COMMACK

NOVA SCIENCE PUBLISHERS
283 Commack Road
Suite 300
Commack, New York 11725

This book is being published under exclusive English Language rights granted to Nova Science Publishers, Inc. by the All-Union Copyright Agency of the USSR (VAAP).

Library of Congress Cataloging-in-Publication Data

Lazernye metody issledovaniĭ defektov v poluprovodnikakh i diélektrikakh. English.
Laser methods of defect investigations in semiconductors and dieletrics.

(Proceedings of the Institute of General Physics of
the Academy of Sciences of the USSR ; v. 4)
Translation of: Lazernye metody issledovaniĭ
defektov v poluprovodnikakh i diėlektrikakh.
Bibliography: p.
Includes index.
1. Semiconductors--Defects. 2. Dielectrics--Defects.
3. Laser beams. I. Manenkov, A. A. II. Title.
III. Series: Trudy Instituta obshcheĭ fiziki.
English ; v. 4.
QC611.6.D4L3913 1988 621.3815'2 88-6622
ISBN 0-941743-15-2

The original Russian-language version of this book
was published by Nauka Publishing House in 1986.

Printed in the United States of America

CONTENTS

Investigation of Impurity Centers in Semiconductors by IR-Laser Emission Scattering Techniques

V.P. Kalinushkin

Abstract: A method is described for investigating "weak" (impurity concentration $\sim 10^{16}$ cm^{-3}) impurity centers: A low-angle scattering technique for IR laser emission which is used to investigate pure silicon and germanium crystals. It is demonstrated that the dominant type of weak impurity clusters in this material includes so-called impurity clouds that are formed from the dissolution of impurity microinclusions embedded in the crystal during crystallization. The process of oxygen cloud interaction with free carriers in Ge is investigated. It is demonstrated that these function as carrier trap centers in γ-radiation detectors fabricated from ultrapure germanium. The influence of impurity clouds on the conductivity of ultrapure semiconductor materials is examined.

INTRODUCTION

Semiconductor electronics has developed at an extraordinarily rapid pace over the last 30 years and has seen significant success and resulted in the development of a number of new fields, including optoelectronics, automation, computer and measurement technology. These achievements have largely been possible due to rapid progress in the fabrication of pure, homogeneous, structurally-complete and relatively inexpensive semiconductor materials. Today, for example, industry manufactures dislocation-free silicon and germanium monocrystals with electrically-active impurity concentrations of $\sim 10^{11}$ cm^{-3} as well as minor dislocation ($N_{Д} \sim 10^4$ cm^{-3}) semiinsulating GaAs crystals and dislocation-free silicon wafers $\sim$ 100 mm in diameter. As a result there have been increasing demands to modernize old techniques and develop new methods of recording and analyzing residual impurities and structural defects in semiconductor crystals.

An area of significant interest in this regard is the investigation of various local inhomogeneities (such as structural defects and impurity clusters), that normally occupy a small portion of the volume in modern semiconductor materials. This is due to the fact that although such inhomogeneities normally have a weak influence on the conduction and measurements of the Hall effect, in a number of cases they may significantly alter both the crystalline properties and the parameters of instruments manufactured using such crystals.

As a result many studies [1-4] have noted that impurity and structural defects have a significant influence on such parameters of semiconductor devices as the breakdown voltage, the bias current in diodes, the E-I characteristics, etc. Another example of the influence of inhomogeneities on semiconductor properties includes large-scale traps that produce such phenomena as negative photoconduction, long-term relaxations and photoelectric fatigue [5, 6].

The existence of impurity clusters in crystals containing a large concentration of random impurities may cause significant changes in crystal properties in the course of industrial heat treatments. In many studies [7-11] the appearance of various impurity levels and recombination centers as a result of thermal treatments has been attributed to the interaction between rapidly diffusing impurities and various structural defects and impurity centers. It is also interesting to investigate the nature of the thermal stability of impurity centers that are produced in semiconductors during gettering [12, 13] which is employed to eliminate point centers from the working side of the substrate.

In this regard the process of obtaining more comprehensive information on all existing structural defects and impurity centers in semiconductor crystals has been a subject of indisputable interest. However the majority of methods of investigating inhomogeneities in semiconductors that were widely used in the mid 1970s allow recording of only structural defects or impurity centers with impurity concentrations of $\geqslant 10^{19}$ cm^{-3}. At the same time comparatively few techniques could be used to investigate relatively "weak" (with impurity concentrations of $\leqslant 10^{16}$ cm^{-3}) impurity centers with dimensions of the order of a few microns and occupying a small ($< 10^{-2}$) portion of the volume of the test sample.

Indeed, the transmission electron microscope and X-ray structural analysis techniques are sensitive to lattice distortions that will be small in the case of the weak impurity inhomogeneities discussed above. The infrared microscope has a low level sensitivity and will only record particles producing attenuation of transmitted light of greater than 0.1%. Local techniques for analyzing the chemical composition of crystalline surfaces have been developed as well; such techniques include auger-spectrometry, microscopic X-ray (electron-probe) analysis, electron-spectroscopic analysis, secondary ion-ion spectroscopy, etc. The use of such methods makes it possible not only to record impurity centers but also to determine the composition of the impurities comprising the centers. However the sensitivity of such techniques in analyzing small areas is not high and normally amounts to only a few atomic percent, i.e., impurity concentrations of $\geqslant 10^{19}$ cm^{-3} with a spatial resolution of the order of tens of microns.

Weak impurity centers may, evidently, be detected by such techniques as selective etching, diffusion resistance and induced current (EBIC). However, first, these methods are used to investigate objects in direct proximity to the sample surface, which requires careful preparation (chemical etching, polishing and in the case of induced current, application of a Schottky barrier), which may complicate the interpretation of experimental results. Second, such techniques (particularly selective etching) do not allow direct determination of the nature of the observed inhomogeneities. Third, the use of such methods (particularly diffusion resistance) is made more difficult when investigating relatively small defects occupying only small areas. Thus, for example, in the case of the weak impurity centers detected in this study in pure silicon the hit probability for a probe 10 μm in diameter (using the diffusion resistance method) is of the order 10^{-3}.

In the early 1980s the nuclear optical polarization method [14, 15] was proposed for investigating weak impurity centers. This technique was used to obtain extensive information on the impurity composition of inhomogeneities in a crystal bulk, as well as information on their dimensions and concentration. This method was used to detect weak impurity centers in boron-doped Si grown by the Czochralski technique (with a free carrier concentration at T = 300 K of $\sim 10^{14}$ cm^{-3}) [16]. However analogous experiments in pure semiconductors encountered serious difficulties related to significant extensions in experimental duration.

The impurities present in weak inhomogeneities may be recorded by means of neutron activation analysis and mass-spectrometric techniques as well as low-temperature luminescence and capacitive methods; they may also appear in such phenomena as long-term current relaxation, the anomalous temperature relation of the Hall effect [17] and charge losses in γ-detectors [18], etc. However it is clear that the development of methods that allow direct experimental investigation of such impurity centers is of indisputable interest.

In the mid 1970s we proposed using long wavelength light scattering [9, 20] by weak impurity centers for such applications. A description of this method, its capabilities and results from its use to investigate pure Si and Ge crystals is the purpose of the present study.

CHAPTER 1

EXPERIMENTAL SET-UP AND TECHNIQUE

1. Elastic Light Scattering: Method For Investigating Inhomogeneities in Semiconductor Materials

Virtually any inhomogeneity in a crystal (whether it be a dislocation, elastic stresses around the dislocation or any other type of crystalline lattice defect, inpurity precipitate or fluctuations in free carrier concentration) has a dielectric constant that varies from the average throughout the crystal. The deviation in the dielectric constant $\Delta\tilde{\varepsilon}$ produces elastic light scattering. Hence by investigating light scattered by crystals it is possible to obtain information on their inhomogeneities. Light scattering has been used to investigate inhomogeneities in a variety of materials: Bulk defects in amino acid salts [21], microcracks in NaCl [22], impurity inclusions in FaAs and GaP [23, 24], electron-hole droplets in germanium [25, 26] and A-defects in silicon [27].

The light scattering observed in these experiments has normally been described by the Rayleigh-Gans approximation [28]. Here the scattering particles will satisfy the following condition [28][1]:

$$\tilde{\varepsilon} \ll \lambda / (2\pi a), \quad (1)$$

where $\tilde{\varepsilon} = \Delta\tilde{\varepsilon}/\varepsilon$ is the relative deviation of the dielectric constant ε of the crystal; λ is the wavelength of the scattered light in the crystal ($\lambda = \lambda_o/\sqrt{\varepsilon}$); α is characteristic particle size.

Inequality (1) in terms of geometric optics means that the scattering angle ($\sim\tilde{\varepsilon}$) is small compared to the diffraction angle ($\lambda/2\pi\alpha$). Clearly we may assume that the inhomogeneities found in modern minor-dislocation and weakly-doped semiconductors satisfy condition (1). Results from investigations of pure semiconductors by IR micro- and flaw detection [30, 31] support this conclusion.

The validity of relation (1) for scattering particles in pure silicon and germanium crystals (this material was investigated most extensively in our experiments) is also confirmed by results of the present study. In the case where the test samples contain a significant quantity of dislocations, twin boundaries, large impurity inclusions (of the order of several microns or larger), precipitate colonies, etc., the Rayleigh-Gans approximation must be used with extreme care in describing the observed scattering. Its accuracy for investigating structurally-incomplete and heavily-doped semiconductors requires additional analysis in each specific case.

In the Rayleigh-Gans approximation the scattering intensity per unit of solid angle is given by the familiar expression [28]

$$I(\theta) = W_0 C L (G^2), \quad (2)$$

where W_o is the initial luminous flux passing through a sample of thickness L; C is the concentration of scattering centers; the complex scattering amplitude $G(\theta, \varphi)$ is expressed as the integral over the volume of a single scattering inhomogeneity, θ, φ are angles determining the direction of scattered light.

In the present study we will investigate scattering in the incident plane (Figure 1) when $\varphi = 0^o$. The form of $G(\theta)$ depends on light polari-

---

[1]The general theory of scattering by any inhomogeneities is significantly more complex and is described in [28, 29].

zation. If the incident flux is polarized in the XZ scattering plane, while the component that is also polarized in the XZ scattering plane is measured in the scattered flux, then[2] [28]

$$G=\frac{kr}{4\pi}\int(\tilde{\varepsilon}_{xx}\cos\theta-\tilde{\varepsilon}_{zz}\sin\theta)\exp(i(\mathbf{k}'-\mathbf{k})\mathbf{r})dxdydz, \tag{3}$$

where k and k' are the wave vectors of the incident and scattered light rays; r is the radius-vector originating from the center of the inhomogeneity.

If $\tilde{\varepsilon}$ is a scalar quantity dependent only on $r = |\mathbf{r}|$, then (3) is reduced to

$$G=\frac{k^2\cos^2\theta}{æ}\int_0^\infty\tilde{\varepsilon}(r)\sin(ær)dr, \tag{4}$$

where $æ = |\mathbf{k}'-\mathbf{k}| = 2k\sin(\theta/2) \approx k\theta.$

Normally we may represent the quantity $\tilde{\varepsilon}(r)$ in the form $\tilde{\varepsilon}_m f(r, a)$, where $\tilde{\varepsilon}_m$ is the maximum deviation, while $f(r, a)$ is the function determining the radial profile of the inhomogeneity. Then the light scattering intensity $I(\theta)$ is expressed in final form

$$I(\theta)/W_0 = Lk^4a^6C\,\varepsilon_m^2(\cos^2\theta)\,\psi(æa), \tag{5}$$

where the dimensionless function $\psi(æ, a)$ determines the nature of the angular dependence of $I(\theta)$;

$$\psi\,(æ, a) = \left[\int_0^\infty f(\xi)\frac{\sin(æ\,a)}{æa}\xi d\xi\right]^2, \tag{6}$$

i.e., by removing the angular dependence of the scattering, we may determine the dimensions and shape of the scattering inhomogeneities. Then by measuring the absolute scattering intensity at a given angle we may find the value of the combination $C\tilde{\varepsilon}^2_m$. If we know the value of C from any independent measurements it is also possible to determine the value of $\tilde{\varepsilon}^2_m$.

If $\tilde{\varepsilon}$ is not a scalar quantity, the calculations become more complex [28, 29]. In this case in order to determine the characteristics of the scattering inhomogeneities we require a series of experiments on the influence of the orientation of the sample with respect to the probe beam on

---

[2] The formulae remain essentially the same in the case of unpolarized light.

light scattering. Generally analogous experiments will always be necessary when no information is available on the symmetry of the scattering inhomogeneities.

Thus, by analyzing light scattering by optical inhomogeneities we may in principle derive information on their shape, size, orientation and on the concentration and value of $\Delta\tilde{\varepsilon}$.

However we should remember that the observed scattering pattern is determined by the scattering of all inhomogeneities in the crystal. Therefore the experimentally-measured scattering intensity is $I = \Sigma_i A_i I_i(a_i, \theta)$, where $I_i(a_i, \theta)$ is the angular scattering intensity distribution of an i-particle (i.e., particles characterized by the profile function f_i, an i-orientation and dimensions a_i); A_i is the scattering intensity of all existing i-particles.

If the sample contains a large quantity of different defects with comparable scattering intensities, but different dimensions, shapes, orientations, etc., the observed pattern will be determined not only by the dimensions of the scattering particles but also by a wide range of other factors, including the size distribution of the particles, their intensities I_i, relative orientation, etc. In this case it will be difficult to obtain quantitative information on the scattering inhomogeneities.

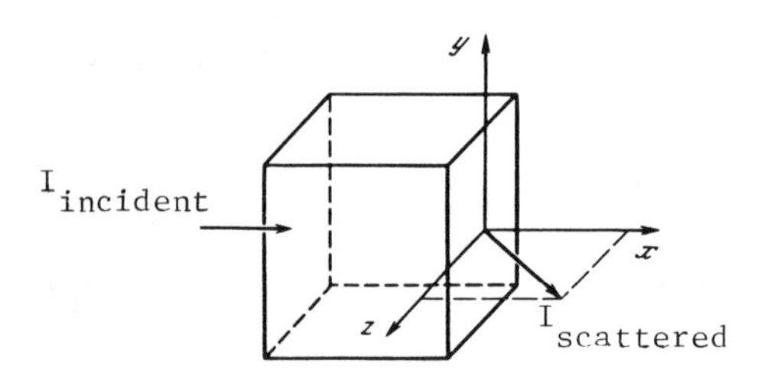

Figure 1. Light scattering scheme.

The situation is enhanced significantly if the crystal contains several particle groups where the spread of particle dimensions within each group is small, while this difference between the sizes of particles from different groups is significantly greater (for example, particles with dimensions of 5-7 and 20-22 μm). It is then possible to classify light scattering by particles of different sizes based on the fact that the larger the scattering particles, the smaller the angles in which the primary mass of scattered light is concentrated. In this case the range of angles in which scattering by particles of one group predominates will be rather significant, which will make it possible to determine the average particle dimension for this group.

Figure 2 provides a scattering diagram of CO_2 laser emission by a pure dislocation-free silicon crystal. Scattering by particles with $a \sim 18$ μm predominates in the range of angles $2° < \theta < 5°$, while scattering by particules with $a \sim 6\text{-}7$ μm is observed when $\theta > 5°$.

It should also be noted that the angular dependence of the scattered light intensity is determined both by the size a and the form of the profile function $f(r, a)$ which is not known in advance. If we know the complete scattering diagram for any given inhomogeneity, we may determine its dimensions and profile function. However in most cases a comparatively small portion of the diagram is measured experimentally. For example, in the present study the scattering diagram was known over a range from 2 to 15°. It is possible to partially eliminate this difficulty by virtue of the fact that it is often possible to establish the form of the profile function by the diagram pattern. For example, Figure 3 gives the angular dependencies of scattering intensity for inhomogeneities with different profile functions. It is clear that their shapes are quite different.

It follows from this discussion that an unambiguous interpretation of the experimental data from light scattering is possible and therefore additional experiments are necessary for a proper interpretation of the results in order to establish the nature of the observed inhomogeneities and for comparison to data using other techniques.

The drawbacks of this technique also include its low information content. Indeed, the information obtained by this technique is reduced to the dimensions, shape and concentration of scattering inhomogeneities. We should, however, note that in a number of cases (which will be discussed below) it is possible to derive a significant volume of information on the nature of the scattering particles by investigating the influence on light scattering of the temperature of the sample, the concentration of nonequilibrium carriers (generated either by photoexcitation or by injection across the p-n-junction) as well as external electric and magnetic fields, the wavelength of the probe light, etc.

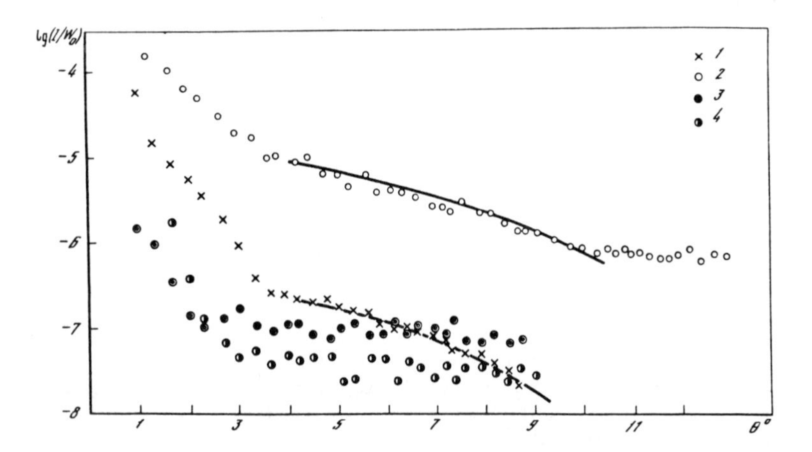

Figure 2. Scattering diagram for group I silicon crystals. 1 - Dislocation silicon; 2 - A-Si; 3 - B-Si; 4 - C-Si.

The advantages of this method include its high sensitivity which is largely the result of the most recent achievements in electronics and laser technology. By using photon counters and an optical master oscillator it is possible to record luminous fluxes from a few to several tens of photons per second in the transition from the visible to the far IR [30, 31].

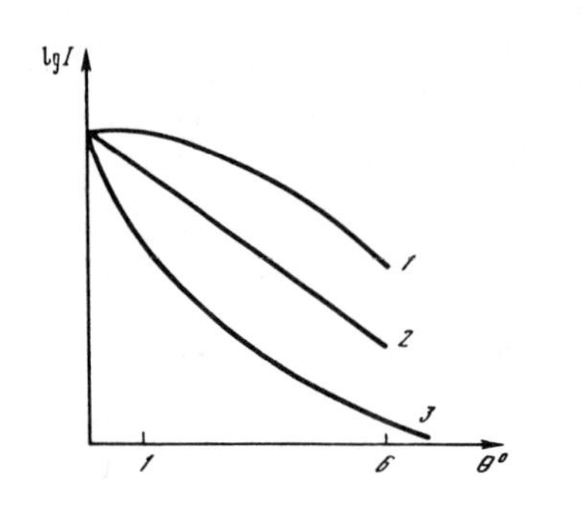

Figure 3. Calculated scattering diagrams for particles with different profile functions.
1 - $f(r, a) = \exp(-r^2/a^2)$;
2 - $f(r, a) = (1 + r^2/a^2)^{-2}$;
3 - $f(r, a) = (1 + r/a)^{-1}$.

An indisputable advantage of the light scattering method (compared, for example, to electron microscopy and local chemical surface analysis techniques) is the possibility for investigating inhomogeneities in the crystal bulk without subjecting the sample to strong external actions. We should also note the comparatively low cost of the experimental equipment required for this method and the rather rapid experimental return and relative simplicity of preparing samples for the procedure.

We will now examine certain interesting features of the light scattering technique that, in our view, make it possible to significantly increase its capabilities for investigating semiconductor inhomogeneities.

1. As indicated by formulae (5), (6) the scattering intensity is $I \sim 1/\lambda^4$ and, therefore, if $\tilde{\varepsilon}$ is independent or has a weak dependence on the wavelength, its value will drop off rapidly with growth of λ. As a rule all lattice defects, elastic stress fields around the defect and neutral impurity centers have a $\tilde{\varepsilon}$ that is virtually independent of λ. If the semiconductor contains regions with an elevated (or reduced) concentration of free carriers, the change in the dielectric constant of these regions is determined by the formula

$$\Delta\tilde{\varepsilon} = -4\pi e^2(n - n_\infty)/m_n c^2 k^2, \tag{7}$$

where e is electron charge; m_n is its effective ohmic mass: n_∞ is the concentration of electrons (or holes) far from the defect; c is the speed of light in a vacuum. In this case $\Delta\tilde{\varepsilon} \sim \lambda^2$.

Since $I \sim \tilde{\varepsilon}^2$ the intensity of scattering by variations in the free carrier concentration will be independent of the wavelength. Hence increasing the wavelength of the scattered light will make the light scattering technique sensitive to the existence of regions with an elevated free carrier concentration in the crystals. For example it will be demonstrated below that light scattering with $\lambda \sim 10$ μm by a cluster of ionized impurities is caused primarily by the change in carrier concentration rather than the change in concentration of the impurity itself. Such regions are formed due to the existence of impurity centers in the semiconductors as well as local variations in the bandwidth, space charge regions around the dislocations, etc.

We note that increasing the wavelength of the scattered light has made it possible to solve one additional important problem: Reducing the influence of light scattering by the surface microrelief, since the scattering intensity by the microrelief will also diminish as $\sim 1/\lambda^4$.

2. Since the majority of the light scattered by inhomogeneities of size a is concentrated within angles of $\theta < \lambda/2\pi a$, then by investigating light scattering at various angles and (or) varying its wavelength λ, we may classify scattering by optical inhomogeneities of different size. For example, if a crystal contains two groups of inhomogeneities with significantly different dimensions ($a' \gg a''$), in order to observe the larger inhomogeneities we must select the intervals for recording the diagram and the scattered light wavelength λ such that they satisfy the condition

$$\theta' < \lambda'/2\pi a'. \tag{8}$$

For recording small particles this condition takes the form

$$\lambda''/2\pi a' \ll \theta'' < \lambda''/2\pi a''. \tag{9}$$

3. It is particularly important to point out the following important fact. As indicated by formula (5) light scattering intensity is determined not only by the concentration of scattering centers C and the magnitude of deviation of their dielectric constant $\tilde{\varepsilon}$, but also is highly dependent (as a^6) on the size of the scattering particles. As a result it is possible to differentiate light scattering by "weak" (with a small value of $\tilde{\varepsilon}$) yet rather large inhomogeneities from scattering by "strong" (with a large value of $\tilde{\varepsilon}$), yet small particles.

This entire discussion reveals that in order to investigate defects whose $\Delta\tilde{\varepsilon}$ value is independent of the wavelength (for example, dislocations, A-defects, impurity inclusions and precipitates), it is most advisable to use probe emission with the lowest possible wavelength. For semiconductors this normally is light with a quantum energy somewhat below the bandgap. In this case the problem of recording scattered light is simplified, both by virtue of the increase in scattering intensity and the possibility for using highly-sensitive detectors (for example, photoelectron multipliers). By varying the range of measurement of the scattering diagram it is possible in principle to observe structural defects and impurity centers of different sizes.

An example of such research includes experiments to investigate light scattering at $\lambda \sim 1$-2 μm in $\theta \sim 90°$ by Si and GaAs crystals [23, 27]. Such an experimental scheme is convenient for recording inhomogeneities with $\Delta\tilde{\varepsilon}$ independent of θ and with dimensions $a \leqslant 1$ μm. Such studies report observing A-defects in Si and impurity inclusions in GaAs; their shape and dimensions were determined.

However from our viewpoint it is more advisable to use such familiar techniques as transmission electron microscopy, structural X-ray analysis and local chemical surface analysis techniques to investigate the defects discussed above. These methods are well-developed and yield more extensive information than the light scattering method. Transmission electron microscopy makes it possible to visualize separate defects and yields information on their nature, while the local chemical analysis technique makes it possible to determine the chemical composition of impurity centers and structural X-ray analysis may be used to investigate elastic stress fields surrounding defects. The prospects for the use of short wavelength light scattering are most commonly related to experimental simplicity, speed and economy and, possibly, the high sensitivity of the technique.

We believe that experiments involving longer wavelength light scattering are of significantly greater interest. Indeed, in this case the technique becomes sensitive to the existence of free carrier clusters in the semiconductors, and the high sensitivity makes it possible to register carrier clusters with concentrations of $n \leqslant 10^{15}$ cm^{-3}. It is difficult to investigate such weak inhomogeneities in view of the absence of methods for recording such inhomogeneities.

There are a number of methods that make it possible to obtain sufficiently extensive information on the nature of the scattering particles if they take the form of free carrier clusters.

1. The use of light scattering at two different (and sufficiently separated) wavelengths makes it possible to establish whether the observed particles are free carrier clusters or inhomogeneities whose $\Delta\tilde{\varepsilon}$ is independent of the wavelength of the probe emission.

2. It is possible to obtain extensive information on the nature of free carrier clusters observed in light scattering from experiments on the influence of sample temperature on light scattering. If these clusters are formed in regions with an elevated impurity concentration due to their ionization (since a change in temperature causes the degree of ionization of the impurities to change correspondingly), the carrier concentration n in the cluster will also change. The intensity of light scattering by the clusters will change as well (approximately as n^2). These experiments make it possible not only to establish the nature of the detected free carrier clusters, but also to determine the ionization energy of the impurities comprising such clusters.

3. In our view it is also interesting to investigate the influence of photoexcitation of free carriers on light scattering.

First such experiments may be used to identify inhomogeneities interacting with free carriers (for example, impurity centers containing recombination centers or traps), and to investigate the interaction mechanism. Indeed, if there exists inhomogeneities that appear as large-scale traps or recombination centers, regions with an elevated (or reduced) carrier concentration will be formed during the generation of the nonequilibrium carriers near such inhomogeneities, which may be registered by means of the light scattering technique.

The pulsed generation of carriers (for example, by pulsed irradiation) is of special interest. Indeed, the scattering intensity for the case of light scattering by inhomogeneities interacting with carriers is attributable to the average carrier concentration in the sample. Hence, in pulsed carrier generation, pulses of light scattered by the inhomogeneities will be observed, which will make it possible to register such inhomogeneities even in the presence of more intense (yet at a constant background level) light scattered by all remaining particles in the sample. By investigating the kinetics of scattered light pulses and the influence of various actions on the light (for example, CW impurity irradiation) will provide extensive information on the carrier/inhomogeneity interaction mechanism.

Second, irradiating samples with light with a quantum energy below the bandgap makes it possible to observe the photoionization of the impurities comprising the clusters. As a result it is possible to observe scattering by clusters of deep (i.e., nonionized) impurities, and, moreover, it is also possible (in conjunction with investigations of the temperature dependence of scattering intensity) to measure the photoionization energies and cross-sections of impurity levels comprising the clusters.

It is also possible, clearly, to obtain additional information on the nature of free carrier clusters observed by means of light scattering by investigating the influence on light scattering of such external factors as electric and magnetic fields, pressure, etc.

It follows from this entire discussion that long wavelength light scattering is an efficient and informative technique for investigating weak free carrier clusters, including weak impurity clusters.

In the present study we have focused primarily on CO_2 laser emission (λ_0 = 10.6 μm) scattering by semiconductor crystals in a range of angles of $1° \leqslant \theta < 15°$. The selection of a CO_2 laser as the probe emission source is determined by the fact that industry is presently manufacturing sufficiently stable CW CO_2lasers and photodetectors sensitive in the 10 μm range. The experimental geometry (the selected range of angles is convenient for recording particles with $a \sim$ 4–20 μm) is the result of the fact that since the dimensions of known structural defects and impurity inclusions and precipitates in pure Si and Ge crystals (it was precisely this material that was the primary focus in the present study) are no greater than 1 μm, it becomes possible to use both the long wavelength of the probe emission and the large size of the weak inhomogeneities to suppress scattering by these particles.

2. Experimental Configuration

The primary difficulty in investigating CO_2 laser emission scattering by weak inhomogeneities is the low scattering intensity, which is significantly below the threshold sensitivity of photodetectors in the range of interest. As noted above one method of recording weak light signals is heterodyning [31]. This method is based on the interference of two monochromatic luminous fluxes (the measured flux and a reference flux) at the photodetector input. The purpose of this section is to consider this method for application to scattered beams and to describe the experimental set-ups employed in the present study.

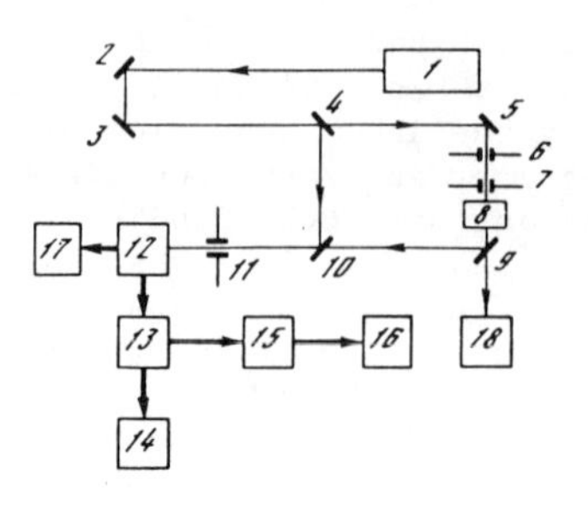

Figure 4. Experimental set-up for investigating small-angle light scattering.
1 - CO_2 laser; 2, 3, 5, 9 - metallic mirrors; 4, 10 - semitransparent splitters; 6, 11 - diaphragms; 7 - non transparent screen; 8 - test sample; 12, 18 - photodetectors; 13 - amplifiers; 14 - oscillograph; 15 - synchronous detector; 16, 17 - electrometers (digital voltmeters).

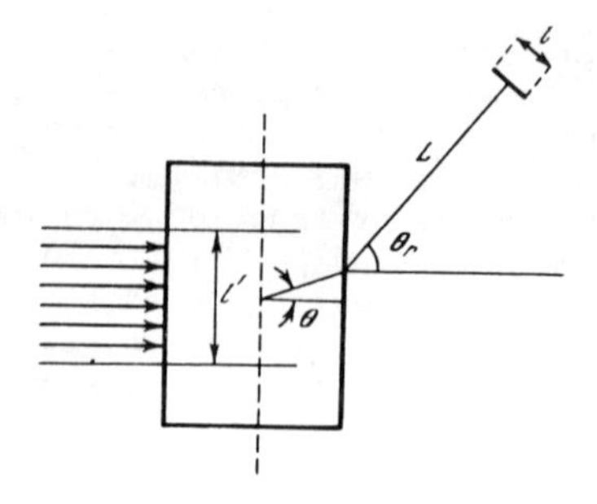

Figure 5. Light scattering scheme for the case of scattering by a single center.

A block diagram of the experimental set-up used in the present study is shown in Figure 4. A high-stability CW CO_2 laser operating at the fundamental TEM_{ooq} mode at a power up to 3 W was used as the scattered light source. Mirrors 2, 3 were used to direct the emission parallel to the plane in which all components of the optical system lie and travel. Semitransparent splitter 4 splits the laser emission into two rays: A reflected (reference) ray and a transmitted (signal) ray. Mirror 5 is used to focus the signal ray on the test sample in the form of a plane-parallel wafer with a diameter of greater than 1.5 cm and a thickness of 0.5 mm; the wafer facets were polished mechanically to optical standards. The sample was mounted behind the diaphragm on a screen that has good absorption at 10.6 μm.

The signal ray is transmitted through the sample and is partially reflected off its surfaces and is scattered by the inhomogeneities within the crystal (attenuation due to initial beam scattering is negligible). After passing through the crystal the beam is focused on the photodetector by means of mirrors 5, 9. The reference and signal rays are merged by means of mirrors 5, 9 and splitters 4, 10; this position of mirror 9 is taken as the coordinate origin (scattering angle $\theta = 0°$).

The angular dependence $I(\theta)$ is read by displacement of mirror 9 perpendicular to the transmitted beam. The scattered light in this case is also the signal ray and is focused on the photodetector and combined with the reference ray by rotation of mirror 9 about the vertical axis. Since mirror 9 is quite large, the scattered and reference rays pass through diaphragm 11 and impact the photodetector. Diaphragm 6 and the nontransparent screen are needed to reduce CO_2 laser emission scattering by mirrors 2, 3, 5 and splitter 4. The range of scattering angles θ has a lower limit imposed by the divergence of the probe ray and an upper limit imposed by the total internal reflection angle (17° for Si and 14.5° for Ge).

Cd-Hg-Te photoconductive cells operating at 77 K were used as the photodetectors; the detection area S was less than 0.05 mm^2. The reference beam power was a few milliwatts for Cd-Hg-Te. The power density in the probe flux (within the crystal) was varied from 1 to 5 W/cm^2. The probe beam diameter was approximately 2-3 mm. Plane-parallel or wedged Ge, NaCl and BaF_2 wafers were used as the splitters.

The signal flux was amplitude modulated mechanically by a modulator. In order to reduce vibrational influences all optical elements in the configuration were mounted on a solid cast-iron support. The signal from the photodetector was injected to an amplifier with a bandwidth of a few tens of hertz to record the a.c. signal component, and was then injected to a digital voltmeter (or electrometer) to determine the d.c. component.

The instantaneous power P at the detector input is equal to

$$P = \frac{cS}{4\pi}\left[E_0\cos\omega t + E_r(t)\cos(\omega t + \varphi)\right]^2, \tag{10}$$

where c and ω are the light velocity and frequency; E_o, E_r (t) is the electric field amplitude for the reference and recording fluxes (the latter is modulated at a low frequency of ω_m); φ is the phase difference between the two fluxes.

Expression (10) is valid if the quantity φ is constant along the receiver area. In fact the fluxes from the various scattering centers may form a small angle of $\sim l'/L$ with the reference ray (Figure 5), where l' is the radius of the illuminated portion of the sample; L is the distance from the sample to the detector. This causes a phase differential of $\Delta\varphi \sim k_o l l'$, where $k_o = 2\pi/\lambda$ is the wave number, l is the area of the receiver surface ($l < l'$). The geometric parameters l', l, L may be selected so that $\Delta\varphi$ is sufficiently small.

The measured power was obtained by averaging (10) over the oscillation period $2\pi/\omega_m$ and consists of the d.c. component P_o (the reference flux power) and the modulated component $P_m(t)$. If $E_r \ll E_o$, then

$$P_m(t) = \frac{cS}{4\pi} E_0 E_r(t)\cos\varphi. \tag{11}$$

The maximum value P_m (corresponding to $\varphi = \pi n$, where n is an integer) is achieved by slight displacement and rotation of mirror (9). In this case (11) will take the form

$$P_m^2 = 4P_0 I_r(t) S L^{-2}, \tag{12}$$

where I_r is the intensity (i.e., the power per unit of solid angle) for the scattered light with the same polarization as the reference ray.

The sensitivity of the receiver system using a master oscillator is determined by the modal structure, as well as the temporal and spatial stability and power of the CO_2 laser as well as tuning accuracy (the convergence of the reference and signal rays at the photodetector input). For our low-frequency modulation and bandwidth of the amplifier circuit the noise limiting the sensitivity of the configuration was determined by laser emission instability ($\geqslant 0.1\%$). In these conditions the minimum value of I_r at which the signal-to-noise ratio is equal to 1 was 10^{-9} - 10^{-10} W/sr.

In experiments devoted to investigating light scattering by inhomogeneities the primary task is determining the dependence on the angle θ of the ratio of the scattered light intensity I_r per unit of solid angle to the incident power I_r/W_o. In order to determine I_r we must know the values of $P_m(t)$ and P_o that may be read from digital amplifier 13 and digital voltmeter 17, respectively. We may determine W_o by using, for example, a calorimeter behind mirror 9. However a Ge:Au photodetector was normally used in this position; this photodetector was precalibrated using a calorimeter. Occasionally this detector was excluded.

Indeed, generally speaking, the probe ray power W_o is proportional to the reference ray intensity $I_{o\pi}$, i.e., $W_o = BP_o$, where B is determined by the reflection coefficient of splitters 4, 10, as well as the characteristics of the photodetector and the experimental geometry. Hence in order to determine I_r/W_o it is sufficient in principle to know the values of $P_m(t)$ and P_o. We should, however, note that it is more accurate to determine W_o by means of detector 18, since the value of the coefficient B may depend on the tuning of the reference ray.

In order to investigate light scattering at low temperatures the sample was placed in a cryostat in a cooling vessel. Liquid nitrogen (77 K) or dry ice (216 K) were used for cooling. Weakly-dispersive silicon windows for which $I/W_o \sim 10^{-6}$ sr^{-1} (where W_o is the incident light ray power) were used to introduce the laser probe emission and extract the scattered light; I for this material was virtually independent of θ. The cryostat design provided a variation of θ from 0 to 8°. The sample was placed in the cryostat and the $I(\theta)$ scattering diagram was recorded at room temperature T_k. Then the cryostat (fastened to the experimental assembly) was filled with an appropriate coolant and the $I(\theta)$ curve was recorded at 77 or 216 K. The coolant was then evaporated off using a special heater and the scattering diagram was again measured at T_k. Such an experimental configuration guaranteed investigating the same crystal area at different temperatures.

The scattering diagrams were also recorded at temperatures above T_k (between 300 and 450 K). For such measurements the sample was placed in a solid metallic housing heated by an electric furnace. A thermocouple was used to monitor temperature.

The present study also investigated the influence of pulsed carrier photoexcitation on light scattering by germanium crystals. In this case, modulation was provided by pulses of light-generated free carriers, i.e., only changes in scattering intensity were recorded and these depended on the concentration of free carriers in the sample. Detector 18 in conjunction with the scattered light pulse were used to record the signal from CO_2 laser absorption by the generated free carriers.

A $CaF_2:Dy^{2+}$ laser was employed for photoexcitation of Ge; this laser produced giant pulses ($\tau \sim 40$ nsec) at a power of 200 kW per pulse with a repetition rate of 400 Hz at 2.36 μm [33]. Photoexcitation was achieved by two-quantum absorption of dysprosium laser emission [34] which made it possible to generate carriers in the crystal bulk. A cylindrical lens was used to create an experimental geometry (Figure 6) that is very convenient for investigating light scattering.

By using a dysprosium laser it is possible to uniformly excite carriers in a 5 x 5 x 5 mm^3 sample up through concentrations of 10^{16} cm^{-3}. At higher concentrations carrier excitation becomes significantly inhomogeneous due to the selective absorption of dysprosium laser emission by the holes. The carrier concentration is determined by the familiar two-photon absorption coefficient [34] and the level of selective absorption of the probe beam [35].

Due to the small relative pulse duration (the lifetime of the free carriers in germanium determined by the decay of the absorption signal was in the order of 200 ~ 300 μsec) and the frequency instability of the laser, the amplifier operated over a broadband range (~ 100 kHz). The signal was taken from the amplifier output and was input to the synchronous detector and then the ED-05M electrometer. The signal level was read on the ED-05M digital instrument. An RC-filter with $1/RC \sim 10^{-4}$ was placed before the amplifier in order to reduce reference ray noise (primarily associated with insufficient power supply stabilization).

The sensitivity of the assembly operating in this mode was $\sim 10^{-7}$ - - 10^{-8} sr^{-1}.

Since scattering modulation in photoexcitation experiments is achieved by the carriers generated in the test sample, there is no need to use a screen or diaphragm. Some experiments devoted to the influence of photoexcitation on light scattering by germanium crystals have been carried out at liquid nitrogen and liquid helium temperatures. In this case the sample was placed in the cryostat described above within a cooling vessel and either liquid helium or liquid nitrogen was then poured into the vessel. The temperature of the samples was not controlled. The silicon windows in this case were replaced by BaF_2 wafers.

We also investigated backscattering (i.e., at 180°) in germanium samples under photoexcitation. The experimental scheme is shown in Figure 7. The CO_2 laser ray impinged upon splitter 6 and the reflected ray was directed back to the detector and was then the reference ray. The transmitted ray impacted mirror 5 and was reflected back to splitter 6. A portion of the ray was reflected off the splitter and impinged upon the sample. The scattered light modulated by the nonequilibrium carrier pulses impacted photodetector 4 after passing through splitter 6. Dysprosium laser 2 was used to generate the carriers. The test samples were cubes (5 x 5 x 5 mm^3) with beveled facets.

The limits on recording the forward scattering diagram (from 2 to 14°) are convenient for investigating particles from 2 to 30 μm in size and by investigating scattering at large angles (for example, 180°) it is possible to register inhomogeneities $\leqslant$ 2.0 μm in size.

The present study experimentally measured the dependence on the scattering angle θ of the modulated instantaneous power component at the input to the detector $P_m(t)$ and determined its absolute value. Then formula (12) was used to determine the value of I_r.

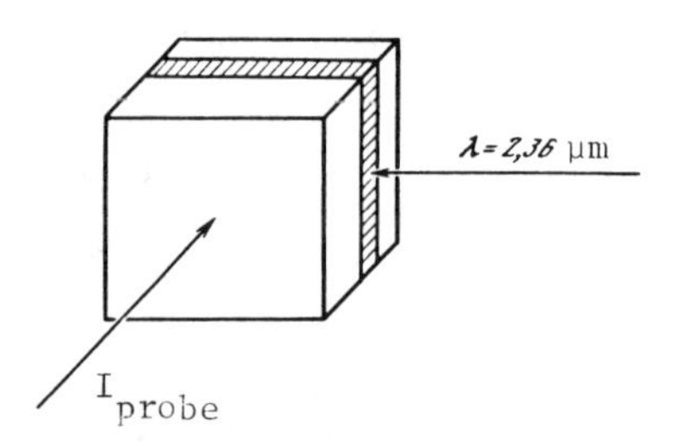

Figure 6. Experimental scheme to investigate the influence of bulk photoexcitation on light scattering.

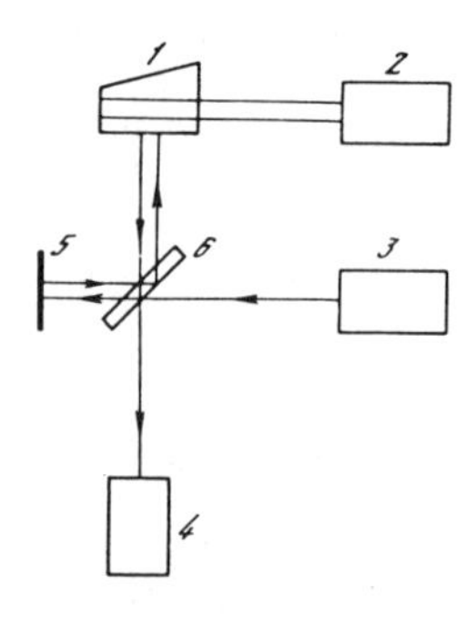

Figure 7. Experimental set-up to investigate light scattering at ~ 180°.
1 - Sample; 2 - $CaF_2:Dy^{2+}$ laser; 3 - CO_2 laser; 4 - photodetector; 5 - mirror; 6 - splitter.

Recording the scattering diagrams on the assembly noted above requires (due to the need to merge the reference and scattered rays) additional tuning of mirror 9 at each point in the diagram. As a result, first, the operator must be better trained and, second, the experimental period is extended. In addition a rather long time period is devoted to plotting of the experimental points and formulation of the diagrams. As a result we designed an assembly that includes a number of improvements (specifically, reversing mirror 9 is replaced by a rotating mirror) and the experimental assembly is hooked up to a computer. The assembly is described in detail in [36]. The sensitivity of this assembly is somewhat lower (10^{-8} - 10^{-9} sr^{-1}), although the diagram plotting time requires only 10-12 minutes and processing of the results is virtually instantaneous.

We investigated pure Si and Ge crystals grown in both a vacuum and in various atmospheres. Detailed characteristics will be given for the samples in the description of test results.

The preparation of the crystals involved slicing plane-parallel wafers from the test material and polishing their facets to optical standards. The wafers had a lower limit on size of 1 cm^2. Their shape was irrelevant. Their thickness normally varied from 0.5 mm to 2 cm.

3. Processing of Experimental Results

As noted above the dependence of I/W_0 on θ is determined experimentally together with the absolute value of the quantity I/W per solid angle unit. The scattering diagram yields information on the size and shape of the scattering particles and the absolute value of I/W_0 determines their concentration and the value of $\Delta\tilde{\varepsilon}$.

We will first determine the value of I_r/W_0. The values of $P_m(t)$, P_0 and P_W were determined by the known E-P characteristics of the detector and the gain of amplifier 13. Then formulae (11), (12) were used to determine the value of I_r/W_0. However, first, we do not know how precisely the reference and signal rays overlap at the photodetector input and, second, the influence of the term $\cos\varphi$ is not clear.

The gain of the master oscillator was therefore determined experimentally as well. The signal level was compared using and excluding the master oscillator. In this case the signal power at the detector input could not exceed fractions of a milliwatt. Therefore either the signal ray passed through filters or light scattered at small angles (up to 1°30') were used as the measured quantity.

The measurements were repeated several times for the same intensity level of the reference beam and then the system was retuned prior to each measurement. We noted that the master oscillator gain for a single intensity of the reference ray could vary by more than a factor of three, while its upper boundary was close to the gain calculated by the reference ray power.

Unfortunately when the transmitted ray was passed through filters the system was detuned. Hence retuning is required after using the filters. As a result the master oscillator gain measured after calibration may differ from the gain for scattered light (by a factor of $\sim$ 3). The small-angle-scattered light gain is measured in direct proximity to the powerful transmitted beam which may significantly distort measurement results. It is not possible to perform measurements at large angles, since it is then not possible to record the signal without the master oscillator.

Therefore we assume that in our experiments the absolute intensity of scattered light was determined with an accuracy in the order of a factor of 3.

The diagrams were normally recorded at points in approximately 30' intervals; characteristic scattering diagrams are shown in Figures 2 and 8. $P_m(t)$, P_0 were measured at each point (if detector 18-P was used) and were used as the basis for determining the value of I_r at the given point.

It is clear from the diagrams that there is a rather large spread of experimental points. There are two reasons for this: The spatial instability of laser emission and the method of recording the diagram. The latter fact is clearly the dominate element. The problem is that each point of the diagram is recorded, as noted above, by additional tuning of mirror 9. Since it is necessary to achieve simultaneous coincidence of the reference and signal rays at the photodetector input as well as a zero phase difference $\Delta\varphi$ of these rays (or a π difference), it is difficult to guarantee that the results obtained at each point in the diagram correspond to precisely the same master oscillator gain. By using a rotating mirror the rays may be combined automatically, although it is not possible to automatically tune to an optimum value of $\Delta\varphi$. Moreover, the spatial distribution of the CO_2 laser emission may change in the course of measuring the diagram, which degrades reference and signal ray matching at the photodetector input and therefore distorts the experimental results. It is not possible to directly monitor ray position during the diagram measurement process in view of their low power.

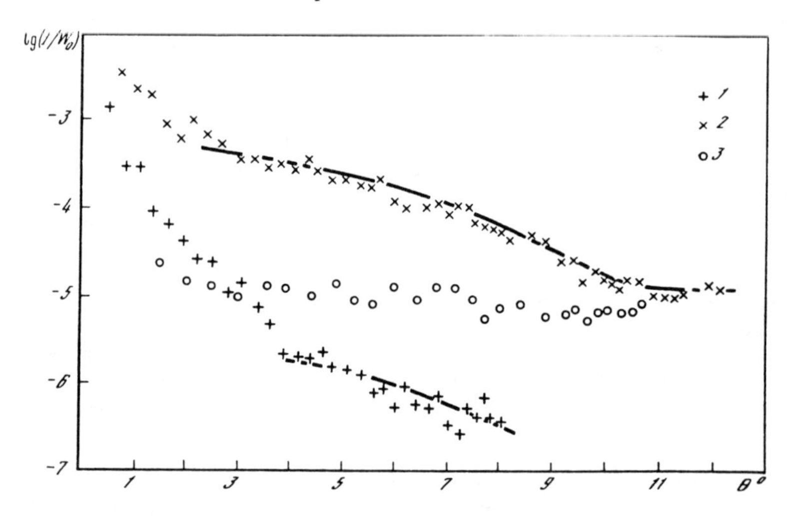

Figure 8. Scattering Diagram For Germanium Crystals.
1 - Dislocation germanium; 2 - A-Ge; 3 - B-Ge.

Since the spread in the experimental data is caused by random variations, the magnitude may be reduced by using broader measurement statistics (as observed experimentally). The experimental data provided here are the average result of several measurements. We note that in order to formulate the diagrams it is not generally necessary to know the absolute values, but rather it is sufficient to measure their relative variations.

Naturally the measurements are performed outside the crystal and hence the true value of the scattering intensity $I(\theta)$ differs from the experimentally observed value due to refraction and reflection off the rear crystal facet. The experimentally observed scattering angle θ_r is related to its true angle θ by the refraction law (Figure 9) $\sin\theta_r = \sqrt{\varepsilon}\sin\theta$. The observed intensity $I_r(\theta)$ is related to the scattering intensity in the crystal bulk $I(\theta)$ by the energy conservation condition and the relation between the solid angles $d\Omega$ and $d\Omega_r$ (before and after refraction): $d\Omega = (\varepsilon\cos\theta/\cos\theta_r)/d\Omega_r$. From here

$$I_r(\theta_r) = (1-j)I(\theta)\cos\theta_r/\varepsilon\cos\theta, \tag{13}$$

where j is the reflection coefficient off the crystal surface which is a known function of θ and θ_r [33]; for example, for light polarized in the *xz* plane,

$$j = \mathrm{tg}^2(\theta_r - \theta)/\mathrm{tg}^2(\theta_r + \theta). \tag{14}$$

We may calculate the desired quantity $I(\theta)$ from formulae (12)-(14) by the measured signal $P_m(t)$; the quantity $I(\theta)$ infers scattering intensity in an infinite crystal. Moreover we must account for the fact that the value of L increases in measuring the diagram:

$$\Delta L = l_0(1/\cos\theta - 1 + \mathrm{tg}\theta), \tag{15}$$

where l_0 is the distance from the sample to the axis along which mirror 9 travels.

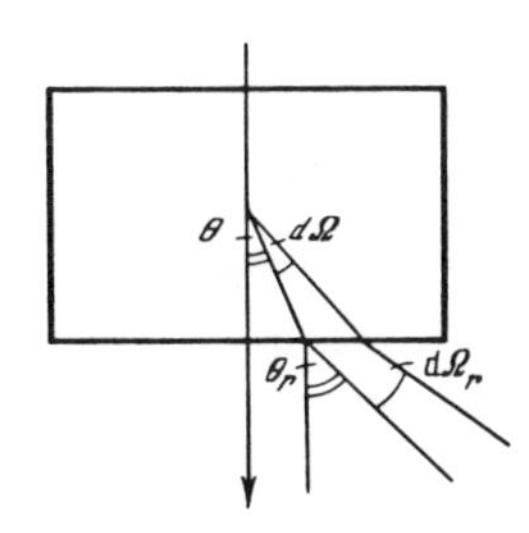

Figure 9. The change in angles at the crystal boundary.

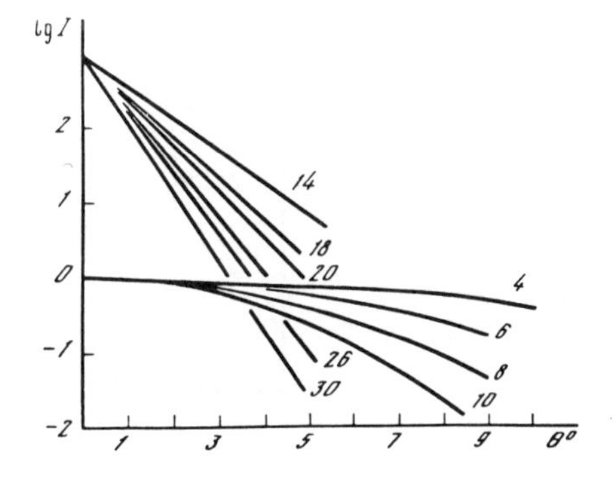

Figure 10. Parametric families of curves for particles of different size (in μm) with different profile functions.

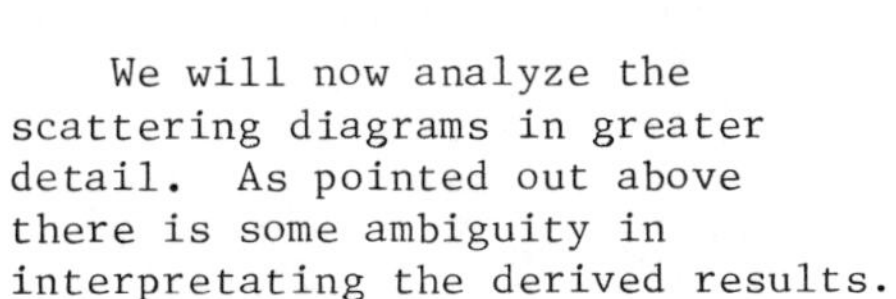

We will now analyze the scattering diagrams in greater detail. As pointed out above there is some ambiguity in interpretating the derived results. We utilized the following scheme for processing results. Normally it is possible to identify regions of characteristic slope and shape in the observed scattering diagrams. Thus, Figures 2 and 8 show a scattering diagram characteristic of the pure Si and Ge crystals tested in the present study.

The following characteristic regions may be identified in the diagrams. The region up to approximately 2° is evidently determined by the divergence of the probe beam. Then the so-called fast part of the angular distribution appears characterized by a strong dependence of I on θ and on the logarithmic scale this dependence is near-linear. This section is absent in certain crystals; up to 5° in sum. Then we observe a region with a smoother dependence of I_{pac} on θ which we will henceforth call the "slow" region. This region is similar to an inverted parabola. The fourth section (in certain crystals the entire scattering diagram consists of this section) is a "plateau", i.e., the scattering intensity in this region has either a weak dependence on θ or is virtually independent of θ.

In analyzing the results we assume that each section of the diagram is determined by scattering of a group of particles of similar size and identical profile distribution function. Parametric families of I(θ) curves calculated by formulae (5), (6) for a single particle (Figure 10) were used to determine these quantities. The experimental curve (corresponding to the diagram section) was combined with the theoretical curves and it was assumed that the group of scattering inhomogeneities is characterized by the parameters of the theoretical curve producing the best agreement.

The fast section of the diagram is accurately approximated by curves for spherical particles with $f(a, r) = (1 + r^2/\alpha^2)^{-2}$ and size a that varied depending on the crystal between 18 and 22 μm. The slow section is accurately described by spherical particle scattering with a Gaussian distribution profile of $f(r, a) = \exp(-r^2/a^2)$ and dimensions a (also dependent on the crystal) from 6 to 9 μm.

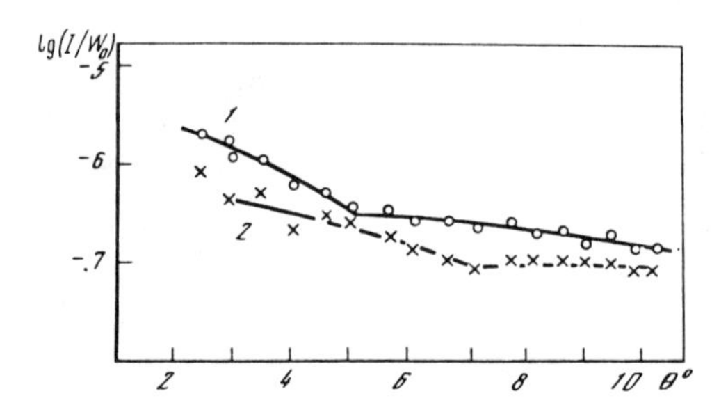

Figure 11. Scattering diagrams of gallium-doped germanium.
1 - Crystal grown in a vacuum; 2 - crystal grown in hydrogen.

The G functions for these inhomogeneities take the form

$$G_1 = \tilde{\varepsilon}_m a_1^3 k^2 \exp(-\text{æ} a_1), \tag{16}$$

$$G_2 = (\sqrt{\pi}/4)\tilde{\varepsilon}_m a^3 k^2 \exp(-k^2 a^2 \theta^2/4). \tag{17}$$

The plateau is characteristic of scattering by particles of small size (< 2 μm). In this case it is impossible to determine either the shape or the exact dimensions of the scattering inhomogeneities.

The present study uses a wide range of additional experiments to determine the nature of inhomogeneities of dimensions $a \sim$ 6-9 μm which confirmed the validity of our approximation. The nature of the "large" inhomogeneities is unknown at present and hence the accuracy of the approximation remains unclear. We believe that it is important to acquire confirmation, since it is often difficult to differentiate particles of different shape by the scattering diagram. It is also rather difficult to differentiate the spherical particles with a Gaussian profile function $f(r, a)$ noted above and sphere identical dimensions in the limiting range of angles. Moreover, it is possible to describe any scattering diagram by introducing a large quantity of various types of scattering centers and by adjusting their size distribution.

The approximation scheme proposed above was used as the initial (zero) approximation. If it was determined from further experiments or from a correlation with results from other methods that our approximation was incorrect, a new approximation of the diagrams was run. Such a situation was evidently employed in the case of semiinsulating GaAs with a high dislocation density [37].

In a number of cases the situation was complicated by the fact that the observed particle groups contained particles of similar size. Thus, Figure 11 gives scattering diagrams for gallium-doped germanium. The approximation scheme proposed above yielded size values for the two particle groups of $a_1 \sim 10$ and $a_2 \sim$ 4-5 μm. In this case it is clearly important to attempt investigating the influence on light scattering by these particles of a variety of actions such as sample temperature, photoexcitation, doping by various impurities, etc. By investigating the dew point influence on light scattering we were able to confirm the existence of three groups of particles of similar size in gallium-doped germanium [38].

In conclusion we note that if the portion of the scattering diagram in which scattering by particles of one type predominates is sufficiently large ($\geqslant 5°$), the accuracy of size measurements is ~ 1 μm. Expanding this interval caused a sharp drop in accuracy and if it fell below 2° it was impossible to determine size.

CHAPTER 2

INVESTIGATION OF LOW-ANGLE LIGHT SCATTERING BY PURE SILICON AND GERMANIUM CRYSTALS

4. Investigation of Pure Silicon and Germanium Monocrystals Having Different Defect Structures

The present study investigated multiple pure silicon and germanium crystals grown by the float-zone method (silicon) and the Czochralski technique (germanium) in both a vaccum and in various atmospheres. In the beginning of this chapter we will focus our discussion on material grown in a vacuum with a low dislocation density or in near-dislocation-free conditions and a concentration of electrically active impurities of $\leqslant 10^{12}$ cm^{-3}. Nonetheless these samples have been characterized by a rather high concentration of impurities detected by neutron activation analysis. Characteristic impurity concentrations for this group of crystals may be found in Table 1.

Results from investigations of pure silicon and germanium crystals grown in various atmospheres and for samples grown in a vacuum yet significantly more pure (see Table 2) will be given in § 8.

The scattering diagrams of group I silicon and germanium crystals are rather similar (see Figures 2, 8). In analyzing these crystals using the scheme described above, we may assume that the samples contain the following groups of inhomogeneities.

1. Spherical particles with $f(r, a) = (1 + r^2/a^2)^{-2}$ and dimensions a varying depending on the crystal between 18 and 22 μm. Scattering by these particles is described by the fast portion of the scattering diagram. Henceforth we will call the particles nonGaussian inhomogeneities. Occasionally nonGaussian inhomogeneities with $a \sim 14$ and 8-9 μm have been observed in group I silicon.

2. Spherical particles with a Gaussian distribution profile $f(r, a) = \exp(-r^2/a^2)$ and dimensions a in the initial crystals and crystals grown in nonspecialized conditions: From 6 to 9 μm. Scattering by these particles accounts for the slow section of the diagram. Henceforth we will call these either Gaussian particles or "6-micron" particles.

3. Inhomogeneities $\leqslant 2$ μm. These correspond to the plateau sections of the diagrams. We will call these the "small" particles.

In order to establish the nature of inhomogeneities detected in the silicon and germanium crystals this section will investigate light scattering by silicon samples with an approximately equal composition of residual impurities yet a different defect structure. The nature of analogous scattering particles in germanium is examined in § 7.

Below we will present results from investigations of dislocation-free silicon crystals grown at a variable growth rate. A change in the growth rate will result in the formation of regions with a variable density of intrinsic microdefects in these samples. At the same time the composition of the residual impurities in these regions will be approximately identical.

Henceforth we will label dislocation-free silicon crystals with A-defects as A-Si, crystals without A-defects with a large concentration of B-defects as B-Si and silicon with C- and D-defects as C-Si.

Analogous to the case of silicon we will label dislocation-free germanium with large dislocation-free defects as A-Ge, and will label dislocation-free germanium with small defects as B-Ge.

Figure 2 gives scattering diagrams for samples sliced from various parts of a pure (concentration of electrically active impurities of $\leqslant 10^{12}$ cm^{-3}) *p*-type silicon crystal grown at a variable rate in a vacuum by the float-zone method. The defect distribution through the crystal blank is shown in Figure 12. The dislocations were introduced into the lower portion of the crystal ($N_{дисл} \sim 10^4 - 10^5$ cm^{-2}) by thermal impact during separation from the melt. The microdefect concentration was determined based on the concentration of etching pits. These measurements were performed at the State Scientific Research and Planning Institute of the Rare Metals Industry. Such measurement results are in agreement with known literature data [39].

The present study has established the nature of Gaussian particles. Therefore we will henceforth focus on research results applying to specifically the inhomogeneities.

As we see from Figure 2, the scattering intensity by 6-micron particles is at a maximum in A-Si crystals. In the dislocation region its intensity is significantly less. Scattering diagrams for B- and C-Si appear as a plateau, i.e., small particle scattering predominates in these crystals.

Table 1

Impurity concentration (in cm^{-3}) in group I Si determined by neutron activation analysis

Impurity	Sample No. 1 ρ = 10 kΩ per cm	Sample No. 2 ρ = 2 kΩ per cm	Sample No. 3 ρ = 1500 kΩ per cm	Sample No. 4 ρ = 10-15 kΩ per cm	Sample No. 5 ρ = 10-15 kΩ per cm
Cu	$1.5 \cdot 10^{13}$	$2 \cdot 10^{13}$	$1 \cdot 10^{13}$	$1 \cdot 10^{13}$	$1 \cdot 10^{13}$
Co	$2.7 \cdot 10^{13}$	$1.6 \cdot 10^{13}$	$1 \cdot 10^{13}$	-	-
Fe	$1.1 \cdot 10^{13}$	$9.2 \cdot 10^{12}$	$2 \cdot 10^{12}$	-	-
Zn	$1.3 \cdot 10^{15}$	$1.8 \cdot 10^{15}$	$2.4 \cdot 10^{14}$	-	-
Ni	$2.5 \cdot 10^{15}$	$1.5 \cdot 10^{15}$	$1 \cdot 10^{15}$	-	-
Cr	$3.9 \cdot 10^{12}$	$1.2 \cdot 10^{12}$	$1 \cdot 10^{12}$	-	-
Mn	$1.6 \cdot 10^{13}$	$4 \cdot 10^{13}$	$5.1 \cdot 10^{12}$	-	-
Sb	$2.5 \cdot 10^{11}$	$7.7 \cdot 10^{11}$	$2 \cdot 10^{11}$	-	-
Sc	$5.8 \cdot 10^{10}$	$7.3 \cdot 10^{11}$	$5.1 \cdot 10^{11}$	-	-
Sm	$2.4 \cdot 10^{10}$	$7.7 \cdot 10^{10}$	$4.3 \cdot 10^{10}$	-	-
K	$3.1 \cdot 10^{15}$	$9.3 \cdot 10^{14}$	$1 \cdot 10^{14}$	$1 \cdot 10^{15}$	$1 \cdot 10^{15}$
Na	$4.2 \cdot 10^{15}$	$3.7 \cdot 10^{15}$	$2.7 \cdot 10^{14}$	$8 \cdot 10^{15}$	$9 \cdot 10^{15}$
Ca	$7 \cdot 10^{17}$	$1.2 \cdot 10^{16}$	$5 \cdot 10^{15}$	$3 \cdot 10^{15}$	$2 \cdot 10^{15}$
Ag	$1.6 \cdot 10^{17}$	$3.5 \cdot 10^{15}$	$3.5 \cdot 10^{15}$	$2 \cdot 10^{15}$	$2 \cdot 10^{15}$

An investigation of the influence of sample orientation on light scattering by Gaussian particles has demonstrated that these inhomogeneities indeed have spherical symmetry. Thus, Figure 13 gives scattering diagrams of an A-Si crystal for various orientations of the crystal with respect to the probe beam. Gaussian particles predominate in the scattering diagram of this crystal. It is clear that the scattering diagrams are virtually identical.

These results were characteristic of the majority of group I *p*-type silicon crystals investigated in the present study. We should note that we examined a large quantity of samples grown at various facilities in both the Soviet Union and abroad.

Table 2

Impurity concentration* (in cm^{-3}) in group II Si determined by neutron activation analysis

Sample number	Gold	Carbon	Oxygen
1	$1.8 \cdot 10^{10}$	$< 2 \cdot 10^{14}$	$1.1 \cdot 10^{15}$
2	$3.9 \cdot 10^{10}$	$4.3 \cdot 10^{15}$	$< 10^{15}$
3	$4.2 \cdot 10^{9}$	$< 2 \cdot 10^{14}$	$< 10^{15}$
4	$2.6 \cdot 10^{10}$	$< 2 \cdot 10^{14}$	$3.7 \cdot 10^{15}$
5	$1.8 \cdot 10^{10}$	$5.4 \cdot 10^{15}$	$2.4 \cdot 10^{16}$
6	$1.4 \cdot 10^{10}$	$1.2 \cdot 10^{16}$	$< 10^{15}$
7	$1.1 \cdot 10^{10}$	$< 2 \cdot 10^{14}$	$< 10^{15}$
8	$4.2 \cdot 10^{10}$	$1.5 \cdot 10^{15}$	$< 10^{15}$
9	$7.1 \cdot 10^{9}$	$6.6 \cdot 10^{15}$	$< 10^{15}$
10	$2.3 \cdot 10^{10}$	$< 2 \cdot 10^{14}$	$< 10^{15}$

*The Cu, Ga, Na, K, As and Fe concentrations were below the detection threshold which was $\sim 10^{14}$ cm^{-3} for Ga and Fe and was $\sim 10^{11}$ cm^{-3} for the remaining elements.

The correlation between the scattering intensity by 6-micron particles and the concentration of A-defects makes it possible to propose that these inhomogeneities are somehow interrelated. However the dimensions of the A-defects determined by electron microscopy normally amount to 1-2 μm [40, 39] which is significantly less than the particles observed by means of light scattering.

Hence one hypothesis was advanced whereby the 6-micron inhomogeneities are the regions surrounding A-defects with altered properties. Then assuming a concentration of scattering regions C equal to the concentration of A-defects, we may estimate the value of $\Delta\varepsilon$ which was 10^{-4} for A-Si crystals (see Table 3).

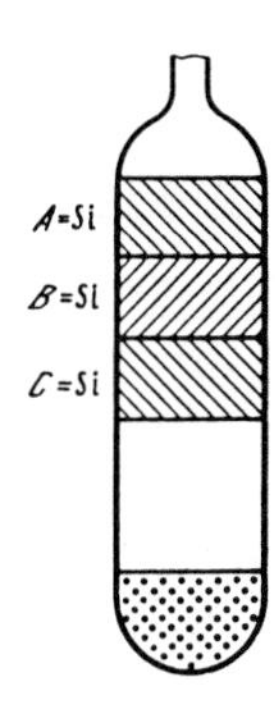

Figure 12. Defect distribution along the silicon ingot grown at a variable growth rate.

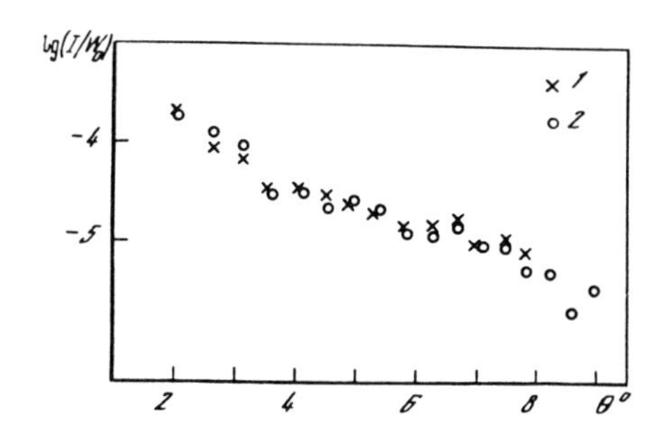

Figure 13. Scattering diagrams for group I A-Si.
1, 2 - Different crystal orientations with respect to the probe beam.

We will now consider the reasons behind the change in the dielectric constant $\Delta\varepsilon \sim 10^{-4}$ at such large ($\sim$ 6 μm) distances. The properties surrounding the microdefect may change for a number of reasons.

1. *Elastic deformation around the microdefect.* This produces a change in $\tilde{\varepsilon}$ itself; second, the deformations change the bandwidth, therefore causing a change in the concentration of free carriers; third, the deformations will interact with impurities causing the formation of a region with an altered impurity concentration around the defect.

The deviation components of the dielectric constants $\tilde{\varepsilon}_{ik}$ resulting from elastic stresses will be linear combinations of the components of deformation $\mathbf{u}_{ik}$, where $\tilde{\varepsilon}_{ik}$ are quantities of the same order as $\mathbf{u}_{ik}$ [41]. We are interested in the region surrounding the microdefect $R \ll r \leqslant a$ (R is the size of the microdefect). In this region $\mathbf{u}_{ik}$ diminishes as r^{-3} [42], particularly for the dislocation loop $u_{ik} \sim bk^2/2r^3$, where $b = 4 \cdot 10^{-3}$ is the Burgers vector.

Substituting the relation $\tilde{\varepsilon}_{ik} \sim r^3$ into (6) we introduce a new integration variable in place of r. The scattering angle θ now enters into only the lower integral limit, i.e., the small parameter $kr\theta$. If the integral over the spherical angles from $\tilde{\varepsilon}_{zx}$ and $\tilde{\varepsilon}_{xx}$ is equal to zero, then integral (6) converges with small $\varkappa r$ and G is generally independent of θ. If the angular integral is not equal to zero, then (6) logarithmically diverges with small $\varkappa r$, and in this case G has a weak (logarithmic) dependence on θ.

Thus, scattering by an elastic deformation yields only a constant background with small θ. Since $\tilde{\varepsilon}_{ik} \sim bk^2/r^3$, integral (6) is equal to bk^2R^3 within an order of magnitude. From here the constant background is $\sim 5 \cdot 10^{-9}$ W/sr, which is significantly less than the experimentally measured value of the scattering intensity.

Table 3

Parameters of scattering inhomogeneities

Substance	Type of defect	$C\tilde{\varepsilon}^2_m$, cm^{-3}	$\lvert\tilde{\varepsilon}_m\rvert$	F	ΔN_m, cm^{-3}	$\Delta\bar{N}$, cm^{-3}
Di	Defects with $a \sim 18.5$ μm	$5.5 \cdot 10^{-3}$	$4 \cdot 10^{-5}$	0.08	$1.5 \cdot 10^{15}$	$1 \cdot 10^{14}$
	Defects with $a \sim 6.4$ μm	0.15	$2 \cdot 10^{-4}$	0.003	$7 \cdot 10^{15}$	$2.5 \cdot 10^{13}$
Ge	Defects with $a \sim 18$ μm	$3.5 \cdot 10^{-3}$	$3.5 \cdot 10^{-5}$	0.07	$7 \cdot 10^{14}$	$5 \cdot 10^{13}$
	Defects with $a \sim 6.4$ μm	0.25	$3 \cdot 10^{-4}$	0.0003	$5.5 \cdot 10^{15}$	$2 \cdot 10^{13}$

The change in the bandgap ΔE_g is proportional to the dislocation u_{ii} (the relative increase in volume) $\Delta E_g/u_{ii}$ = 4 eV for Si and Ge [41]. For a dislocation boundary loop $u_{ii} \sim bk^2/2r^3$ [42] (on the average within angles of $u_{ii} = 0$). When $r \sim a \sim 6$ μm the dislocation is $u_{ii} \sim 10^{-6}$, i.e., band displacement is insignificant and has virtually no influence on the carrier concentration.

The interaction energy of elastic stresses with the impurity is $w = iKVu_{ii}$ [42] where $K \sim 10^7$ N/cm^2 is the bulk modulus of elasticity, V is the increase in crystal volume from the introduction of a single impurity atom. Taking $V \sim 10^{-23}$ cm^3 we obtain $w \sim 10^{-5}$ eV (at $a \sim 6$ μm), i.e., elastic interaction has no influence on the impurity distribution over distances of interest to us.

2. *Space charge.* If the charge of the microdefect Q has a sign opposite that of the majority carriers, then Q is screened by the majority carriers over a distance of $\leqslant l$ (the Debye length). Since $l \leqslant 1$ μm in the test crystals, such an inhomogeneity will produce only a constant background I. If Q has the same sign as the majority carriers, it is screened by the charged impurities (whose concentration is equal to N_∞). The screening radius in this case is also determined by the total neutrality condition

$$a/l = (4\pi/3)a^3 N_\infty. \tag{18}$$

This formula is valid as long as the potential difference Q/ε_r is sufficiently small (otherwise the charge Q will be screened by the minority carriers). The qualitative distortion to the bands $|\varepsilon Q|/\varepsilon R$ will be less than the bandgap E_g, and therefore $a^3 N_\infty \leqslant \varepsilon E/4l^2$. $a \leqslant 10$ μm for large microdefects ($R \sim 1$ μm) and a high-resistance semiconductor ($N_\infty \sim 5 \cdot 10^{12}$ cm^{-3}) which is comparable to the size of the observed inhomogeneity. However the deviation of the carrier concentration in this model is equal to N_∞, and such a small change in carrier concentration will not yield the desired values of $\tilde{\varepsilon}_m$.

3. *Change in impurity concentration near a microdefect or an impurity inhomogeneity.* We will examine in somewhat greater detail light scattering by a change in impurity concentration. The deviation in impurity concentration may be accompanied by a deviation in the free carrier concentration. Hence $\tilde{\varepsilon}$ is decomposed into the contribution of the impurity itself $\tilde{\varepsilon}_{\text{пр}}$ and the electron $\tilde{\varepsilon}_{\text{эл}}$ and hole $\tilde{\varepsilon}_{\text{дыр}}$ contributions. The electron concentration distribution $n(r)$ and hole concentration distribution $p(r)$ depend on the ratio of the size a and Debye screening length l. If the carrier concentration is $\sim 10^{12}$ cm^{-3}, then $l \sim 1$ μm, i.e., $l \ll a$. In this case the neutrality condition is satisfied (approximately) for each volume element of the inhomogeneity. Specifically, if all impurity atoms have an identical charge, the majority carrier concentration is identical to the concentration of the charged impurity $N^+(r)$.

The electron contribution to $\tilde{\varepsilon}$ for $\omega \ll \omega_c$ (this condition is satisfied for the entire IR range) is determined by formula (7). It may be represented as $j_n(n - n_\infty)$, where j_n is a certain coefficient dependent on the properties of the semiconductor and the wavelength of the scattered light.

The coefficient j_n for light at $\lambda = 10.6$ μm for silicon is equal to $\sim 3 \cdot 10^{-20}$ cm^{-3} and for germanium is equal to $\sim 5 \cdot 10^{-20}$ cm^{-3}. The hole contribution is analogous to the electron contribution, and the value of the coefficient j_p is within an order of magnitude of j_n.

We will estimate the contribution of the impurity itself to $\tilde{\varepsilon}$. Within an order of magnitude $\tilde{\varepsilon}_{\text{прим}}$ is equal to the atomic percent of the impurity $N_{\text{пр}}/N_0$ (N_0 is the concentration of lattice points in the crystal). This estimate is valid for both small inclusions and for the impurity dissolved in the crystal. Thus, in the general case of impurity inhomogeneity $N(r)$ we have (accounting for the background concentration far from the microdefect)

$$\tilde{\varepsilon}_{\text{прим}} = j(N - N_\infty), \qquad (19)$$

where $j \sim N_0^{-1} \sim 2 \cdot 10^{-23}$, which is three orders of magnitude smaller than j_n and j_p, and hence the charged impurity influences $\tilde{\varepsilon}$ primarily through the free carrier concentration.

If the observed 6-micron inhomogeneities are clusters of electrically-active impurities, then we may determine the change in impurity concentration $\Delta N_m = N_m - N_\infty$ by the known value of $\tilde{\varepsilon}_m$ and formula (7), where N_m is the maximum concentration in the cluster and N_∞ is the average impurity concentration in the sample. The total number of charged impurities in the impurity clusters is equal to $-F\Delta N_m$ within an order to magnitude, where $F = (4/3)\pi a^3$ C. The corresponding values of these quantities for A-Ge and A-Si are given in Table 3. These values jump by three orders of magnitude in the case of a neutral impurity cluster.

In principle three models that will produce a change in impurity concentration in the vicinity of an A-defect are possible.

1. *The defect as an escape for impurities.* In this case a region with a lower impurity concentration is formed at significant distances from the defect. The maximum deviation in impurity concentration N^+_m does not exceed N_∞, i.e., for a charged impurity $N^+_m \leqslant 10^{12}$, while for a neutral impurity $N^+_m \leqslant 10^{15}$ cm^{-3}, which is significantly lower than the values given in Table 3.

2. *The defect as a source of impurities.* During crystal cooling (from the crystallization temperature to room temperature) the defect apparently changes its state in several jumps [38, 39]. During this irregular transformation a previous impurity may be released. A region rich in the impurity may be formed around the defect as a result of the diffusion of the impurity.

For example [39] has proposed the following model of A-defect formation. A dislocation loop with a packing defect adsorbing various impurities is first formed from small predecessor-defects (evidently B-defects). As the temperature drops the packing defect vanishes and the impurity adsorbed in the defect is released and begins to diffuse into the sample bulk. We should note that this model has not been confirmed by either calculations or experimental data. We employed this model as a working hypothesis to explain the formation around the defect of a region with an elevated impurity concentration with a size significantly exceeding defect size. The impurity distribution around the microdefect formed from diffusion will take the form

$$N - N_\infty = \Delta N_m \, e^{-r^2/a^2}, \tag{20}$$

$\Delta N_m = M(a\sqrt{\pi})^{-3}$; the characteristic diffusion length is determined by the relation $a^2 = 4\int_{T_f} D dt$, where D is the diffusion coefficient (the integration process is time integration). The quantity D rapidly decays with a drop in temperature T as $e^{-(E/k_0 T)}$, where k_o is Boltzmann's constant; E is the migration energy. Hence $\int D dt$ is calculated approximately and yields

$$a^2 = 4 D_f k_0 T_f^2 / Eq, \tag{21}$$

where $q = -dT/dt$ is the crystal cooling rate (a typical value is $q = 0.7$ K/s). The quantities T_f, D_f, q refer to the onset of diffusion.

Approximately $5 \cdot 10^{13}$ cm^{-3} electrically active impurities must be concentrated to explain the observed absolute scattering intensities in the impurity clouds. This value does not contradict results from radiation activation analysis nor the values given in Table 1. If we assume that the observed scattering results from neutral clouds, then the value of $F\Delta N_m$ will be in the order of $5 \cdot 10^{16}$ cm^{-3} which significantly exceeds the oxygen and hydrogen concentrations in the test samples ($\sim 10^{15}$ cm^{-3}). We should note that the experimentally observed Gaussian profile of the relative deviation of the dielectric constant $\tilde{\varepsilon}$ agrees with the impurity distribution profile in the cloud.

3. *Precipitation of the impurity from a solid solution in the vicinity of the microdefect* (but not in the microdefect itself): A "cloud" of hyperfine precipitates. Such clouds have been observed in the vicinity of dislocations [43] and the possibility exists that an analogous situation will also exist in the vicinity of defects. The nature and mechanism of this phenomenon are unknown and hence it is difficult to analyze, although in principle the result will be analogous to the defect acting as a source of impurities: Enriching the region around the microdefect with the impurity.

Thus, of all the factors discussed above only the situation where an elevated concentration of electrically active impurities is formed around the microdefect explains the derived experimental results.

It was possible to confirm this hypothesis by employing experiments on the influence of sample temperature on light scattering.

p-type A-Si crystals (hole concentration of $\sim 10^{12}$ cm^{-3}) grown by the float-zone technique in a vacuum were used for the study. At room temperature the I (θ) curve in a range of scattering angles θ from 2 to 8° corresponded to scattering by Gaussian inhomogeneities with $a \sim$ 7-8 μm, while when $\theta > 8°$ the curve became a plateau.

Figure 14 gives the scattering diagrams of this sample for T = 300 (prior to cooling), 220 and 80 K. After returning to room temperature the scattering pattern is identical to the initial pattern. At 220 and 77 K the scattering intensity is near constant (and is close to the plateau value observed at room temperature in the $\theta > 8°$ range). This means that the scattering intensity by the initial inhomogeneities (7 μm in size) drops off significantly - at least an order of magnitude - with reduction of T. An increase in temperature results in total restoration of the scattering diagram.

The derived results make it possible to conclude that the Gaussian inhomogeneities observed in pure silicon are indeed free carrier clusters that arise due to ionization. With a reduction in temperature the degree of ionization of the centers drops, which produces a significant reduction in free carrier concentration in these regions and at the same time causes a drop in scattering intensity. With T increased to 300 K the degree of ionization and the scattering intensity are restored. If the scattering inhomogeneities are not clusters of impurity centers (but rather are intrinsic or impurity structural defects), the temperature dependence will be weak.

We may obtain additional information on the nature of the observed impurity clusters by investigating the influence of various thermal annealing states on light scattering.

5. The Influence of High-Temperature Heat Treatments on Light Scattering by *p*-Type Silicon

In order to test our hypothesis that the detected impurity clusters are regions of elevated concentration of the dissolved impurity formed from diffusion from some source (an A-defect) we may propose the following experiment: During high-temperature heat treatments there will be diffusion of the impurity comprising this region, thereby increasing its dimensions in accordance with the formula

$$b^2 = b_0^2 + 4D\tau, \tag{22}$$

where b_0 are the initial dimensions; D is the diffusion coefficient of the neutral impurity at the heat treatment temperature; τ is the heat treatment time. The dimensions of the scattering particles will expand analogously.

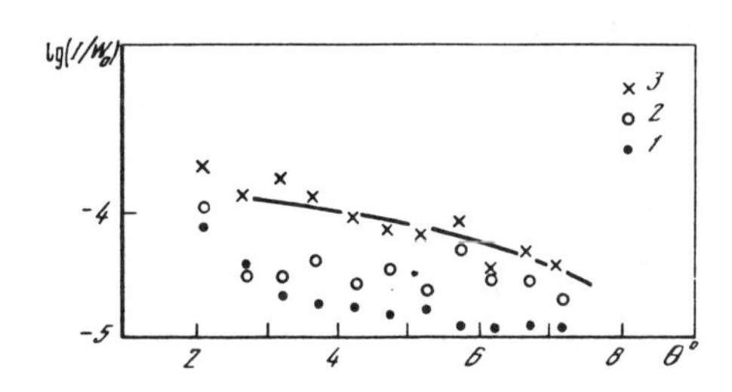

Figure 14. Scattering diagram for group I A-Si.

1 - T ~ 80 K; 2 - T ~ 220 K; 3 - T ~ ~ 300 K.

From this experiment we may also use the diffusion coefficient D to determine the impurity forming the test inhomogeneity.

This section provides results from an investigation of domestic *p*-type pure dislocation-free silicon grown by the float-zone technique in a vacuum with an A-defect concentration of $\sim 10^6 - 10^7$ cm^{-3}. These samples were selected since the scattering intensity by 6-micron particles was the highest in these samples.

The present study also investigated the influence of low-temperature (200°C) and high-temperature (1000° - 1200°C) heat treatments. Low-temperature heat treatments were carried out in air while high-temperature treatments were performed in a special assembly in a variety of gas media. The samples were cooled either in conjunction with a furnace (slow cooling) or by air quenching. The samples were repolished after each high-temperature heat treatment process.

The following was determined from the investigation of the influence of thermal annealing of *p*-type silicon with A-defects (A-Si) at T = 1000°C on light scattering by the silicon.

1. Light scattering corresponding to the slow portion of the diagram vanished immediately after heat treatments followed by quenching and then was gradually restored over a period of approximately 60 days (Figure 15). Heating at 200°C sharply accelerated this process: The restoration occurred over a few hours.

2. An increase in the dimensions of the scattering inhomogeneities as a function of heat treatment time corresponding to the diffusion law (Figure 16) was observed. The diffusion coefficient equal to $1.2\text{-}1.6 \cdot 10^{-11}$ cm^2/sec determined by formula (22) was in good agreement with the known diffusion coefficient of oxygen in silicon of $1.4 \cdot 10^{-11}$ cm^2/sec [44, 45].

3. Only an expansion of size was observed as a result of heat treatments accompanied by slow cooling.

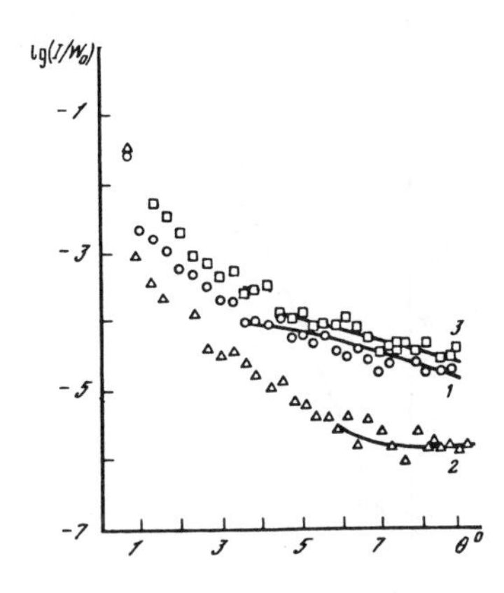

Figure 15. Scattering diagrams of group I *p*-type A-Si.

1 - Initial sample; 2 - heat-treated sample at T = 1000°C for 30 minutes with immediate quenching following the heat treatment process; 3 - heat-treated sample after a 60 day delay.

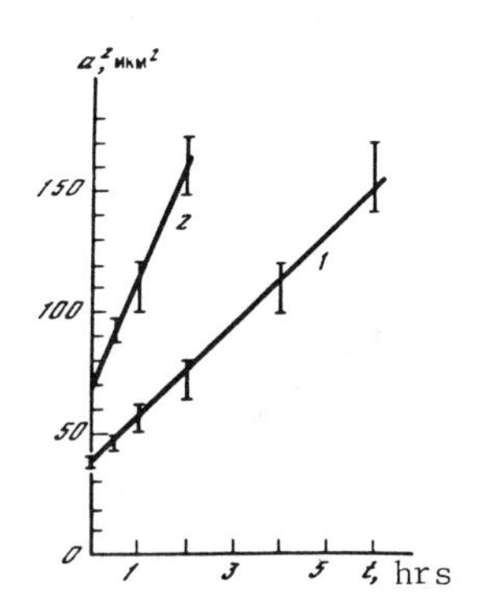

Figure 16. Dimensions of scattering inhomogeneities a as a function of annealing time t (annealing temperature of 1000°C).

1 - *p*-type A-Si; 2 - *n*-type A-Si.

Analogous results were obtained from heat treatments at 1050 and 1100°C, which made it possible to determine the temperature dependence of the diffusion coefficient. This dependence is given in Figure 17 and coincides with the temperature dependence of the diffusion coefficient of oxygen [44, 45].

These experiments made it possible to determine the nature of the detected impurity clusters. Henceforth we will call such clusters impurity "clouds". The clouds observed in silicon are formed by oxygen. In silicon and germanium oxygen is the neutral impurity. As noted above the cluster of neutral impurities weakly scatters light at ~ 10 μm. However at low (near room) temperatures oxygen may form electrically active complexes by interacting with the rapidly diffusing impurities (for example, alkali metals and copper). The oxygen concentration in the cloud is significantly higher than the bulk average concentration and therefore the probability of formation of such complexes will be much greater in the cloud.

As a result a region of elevated free carrier concentration is formed within the cloud and is observed experimentally. We will call this process cloud activation.

During the heat treatment process the complexes break down, the rapidly diffusing impurities escape from the cloud and cloud scattering vanishes.

During the delay at room temperature the rapidly diffusing impurities re-enter the cloud due to diffusion, reform complexes and scattering is gradually restored. Heating accelerates this process while slow cooling causes the impurities to return to the cloud during the cooling process.

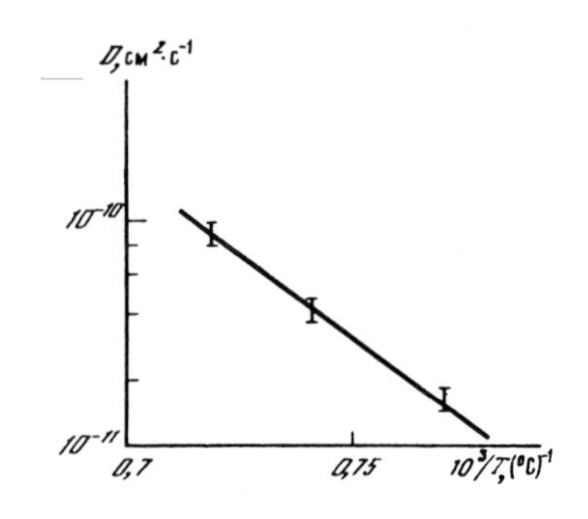

Figure 17. Temperature dependence of the diffusion coefficient of oxygen.

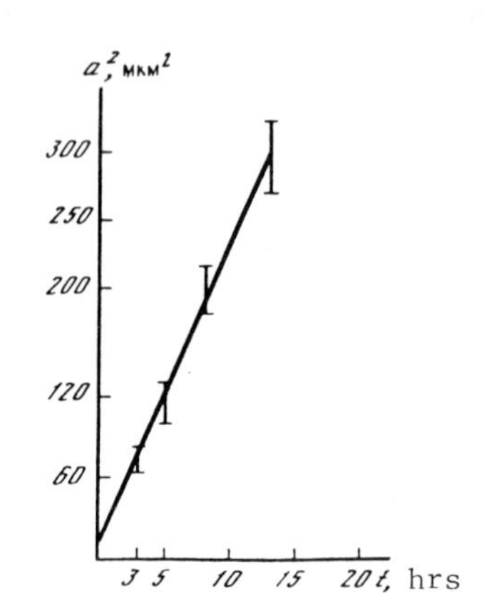

Figure 18. Dimensions of the new optical inhomogeneities a as a function of annealing time t (annealing temperature: 1150°C).

In addition to dissociation of the complexes the heat treatment process is accompanied by oxygen diffusion and an increase (consistent with formula (22)) in cloud size.

Annealing at 1150°C for several hours will cause the oxygen clouds to expand significantly and the corresponding segment of the scattering diagram will shift to the range $\theta \leqslant 2°$. In the initial samples strong scattering by large inhomogeneities of ~ 20 μm were observed at such angles (see Figure 2). In the range $\theta > 2°$ new inhomogeneities were observed in the samples also corresponding to a Gaussian profile. Their size increases over time (Figure 18) in accordance with equation (22) where the initial dimensions are $a_0 \sim 3$ μm with a diffusion coefficient $D = (1.6 \pm 0.3) \cdot 10^{-11}$ cm²/sec. Only carbon ($2 \cdot 10^{-11}$ cm²/sec) has such a value among known impurities [44, 45].

Therefore in addition to oxygen clouds silicon crystals (at least after heat treatments) contain analogous carbon clouds. Again the scattering is not caused by the carbon itself but rather by free carrier clusters (resulting from the formation of electrically-active carbon complexes with rapidly diffusing impurities). Indeed, repeated annealing and quenching strongly reduces scattering by carbon clouds; after heating to 200°C scattering is restored.

In samples not subjected to heat treatments to T = 1150°C the carbon clouds are small ($\leqslant 3$ μm), and their contribution to scattering intensity is virtually constant when $\theta \leqslant 10°$. It is possible that the plateau in the right half of the scattering diagram (see Figure 2) is caused by carbon clouds.

In running these experiments it is important to focus on the following facts: The observed dimensions a (characterizing the free carrier distribution in the cluster $n(r)$) may in the general case differ from b. For definiteness we will consider the case of one type of single-charged donor complexes with a total concentration of $N_1(r)$. The concentrations of the unbound rapidly-diffusing impurity and the phonon acceptors and donors are assumed to be small compared to $n(r)$.

If the complexes form at room temperature we may ignore their dissociation. The rate of formation of N_1 complexes is proportional to the concentration of the unbound neutral impurity $N-N_1$. We will first assume that N_1 is independent of the electron concentration $n(r)$. Then when $N(r) \gg N_1$ $N(r)$ will be proportional to $n(r)$. The quasineutrality condition is satisfied in the inhomogeneous region due to its large size. The electron distribution is simplest in the two limiting cases:

1) The donor complexes are almost completely ionized, i.e. $n = N_1$. Then $n(r)$ is proportional to $N(r)$ and a and b coincide;

2) the donor complexes are weakly ionized, so $n = \sqrt{N_1 n_1}$ (here n_1 is the state density in the conduction band reduced to the energy level of the complexes). Then $n(r)$ is proportional to $\sqrt{N(r)}$. As a result $a = b\sqrt{2}$.

In principle $\dot{N}_1$ may be dependent on n. For example, if the reaction passes through an intermediate positive complex, then $\dot{N}_1$ is inversely proportional to n (since the concentration of such complexes is proportional to $(N-\dot{N}_1)/n$ by the law of mass action). Taking $N_1 \ll N$ we find that $a = b\sqrt{2}$ (with total ionization of the complexes) and $a = b/\sqrt{3}$ (with weak ionization).

In the general case the value of a differs from b by the factor $f \leqslant 2$. If f is not changed from heat treatments, expressions (21) and (22) are valid for a_o and a if we replace D^* by the effective value Df^2.

Measuring a by the low-angle scattering diagram may be considered a unique method of determining the diffusion coefficient of oxygen and carbon (with $f^2 \sim 2$ accuracy).

The good correlation of diffusion coefficients of oxygen and carbon determined by cloud spreading indicates that in the test crystals the interaction between the activation impurity and the oxygen and carbon results in the formation of completely ionized complexes, while their rate of formation is independent of the carrier concentration in the cloud.

If we know which neutral impurities compose the clouds, we may use formula (21) to determine the temperature at which their formation begins. It turns out that oxygen and carbon are liberated at approximately the same temperature of 1300°C, which is significantly below the melting point of silicon.

In concluding this section we note that the scattering diagrams for crystals heat treated at 1150°C reveals scattering by particles with

$a \sim 6$ μm similar to the initial oxygen clouds. It is possible that a certain "trace" remains at the location of the initial oxygen clouds. We have not yet succeeded in determining its nature.

6. Oxygen and Carbon Clouds in Defect-Free Silicon Crystals. Cloud Formation Mechanism

In the preceding sections we assumed that the observed clouds formed from the release of an impurity by an A-defect or its predecessors. This assumption was based on the fact that the scattering intensity by 6-micron initial oxygen clouds was at a maximum in A-Si crystals. However zero or weak scattering by clouds may be caused not only by the absence of the clouds themselves but also by the fact that their activating impurities may be trapped at some other location. For example, they may be trapped by small predecessor-defects in B- and C-Si. The relative weak scattering intensity by dislocation crystals may be attributed to the partial trapping of the activation impurity by the dislocations.

We know [46] that structural changes occur from heat treatments at $T \geqslant 1000°C$ in B- and C-Si. Small etching pits corresponding to the predecessor defects vanish, and a swirl etching picture characteristic of A-Si appears.

Figure 19 shows the change in the scattering diagram of C-Si grown in a vacuum at a growth rate of 6 mm/min from heat treatments at T = 1000°C for 1 hour with subsequent quenching and reheating to 200°C for several hours. It is clear that scattering with $a \sim 6$ μm characteristic of clouds appears. Experiments on the influence of heat treatments on scattering inhomogeneity size analogous to those described in the preceding paragraph have demonstrated that the resulting inhomogeneities are oxygen clouds. The cloud size calculated by formula (22) in the initial (prior to heat treatments) crystal is equal to 4.5 μm. Accounting for the fact that the cooling rate of crystals grown at a high growth rate is significantly greater, we obtain a good agreement with formula (21). Indeed, cloud size is ~ 6 μm in crystals grown at a growth rate of ~ 2.5 mm/min, while cloud size is ~ 4.5 μm in crystals grown at a rate of ~ 6 mm/min.

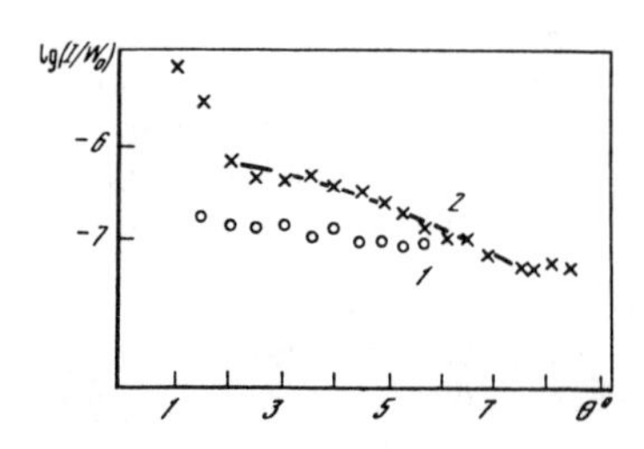

Figure 19. Scattering diagram of group I C-Si.

1 - Initial sample; 2 - after annealing at 1000°C for 1 hour.

Thus oxygen clouds also are found in defect-free crystals. However their activation impurity is evidently concentrated in the minor C- and D-defects. During heat treatments these defects break down and therefore the rapidly-diffusing impurity is liberated and in turn activates the oxygen clouds. Heat treatments of "defect-free" crystals at 1150°C has demonstrated that they also have carbon clouds. We should also note that carbon clouds have been observed in D- and C-Si crystals grown in an argon atmosphere and in a number of B-Si crystals with a lower impurity concentration (see [45]).

Consequently the hypothesis of cloud formation as a result of A-defect release of the impurity is not likely to be correct. Hence we proposed a different model of cloud formation that predicts the dissolution of impurity inclusions and the subsequent diffusion of the liberated impurity from these inclusions. The impurity distribution resulting from this process will be described by a Gaussian profile with a characteristic dimension determined by formula (20), where T_f corresponds to the temperature at which the rate of dissolution of the inclusion is maximized.

We will now consider in greater detail both cloud formation possibilities.

1. During cooling from the crystallization temperature T_m to the escape temperature T_f the dissolved oxygen is absorbed by certain microdefects. At T_f these microdefects are converted into another form, thereby causing the liberation of the accumulated oxygen.

2. The oxygen sources are oxide particles trapped from the melt. These particles are dissolved in conjunction with the absorption of interstitial silicon atoms (occupying the area of a particle) or vacancy formation. Hence the rate of dissolution depends on whether or not the crystal contains microdefects (an escape for the vacancies and the interstitial atoms). In principle the rate of dissolution may be maximized at T_f from the transformation.

In both cases the oxygen sources appear due to defect formation processes in the cooled crystal. At present we only know the final result of these processes in dislocation-free silicon: a) at moderate cooling rates ($q \sim 1$ K/sec) so-called A-defects are observed; these are implanted dislocation loops ~ 1-2 μm in size; b) with large q a stratified distribution of small B-defects is observed with $R \sim 0.05$ μm [47, 48]; c) a further increase in q results in a uniform distribution of even smaller defects [49]: C-defects.

One possible high-temperature transformation scheme consists of the successive transitions $C \to B \to A$ (each of these defects may exist in somewhat different forms [38, 39], which additionally increases the number of possible successive transformations). However it is also possible that all observed A-B-C-defects will be absent at high temperatures, and will only appear later after the precipitation of excess impurities (and will appear as parallel competing forms of precipitates).

Information on the microdefects makes it possible to refine the two oxygen source models discussed above.

1. *Preaccumulation of oxygen.* One of the $C \to B \to A$ transformations will occur at T_f. Since the A-defects are absent when $T \geqslant 1050$°C [50] and, moreover, the clouds will also be found in defect-free crystals, the

oxygen-absorbing microdefects may be type C or type B. We will produce a top-level estimate of the number of oxygen atoms M that may be accumulated by a single microdefect between T_m and T_f. The rate of accumulation of M will not exceed the diffusion flow $4\pi RDN_0$, where R is the radius of the microdefect and N_0 is the concentration of the dissolved oxygen. Carrying out time integration of the flow and using (19), we obtain

$$M < \pi R N b_m, \tag{23}$$

where b_m is given by expression (21) with T_f replaced by T_m. Substituting $R = 0.05$ μm into (23) [47, 48], $N_0 = 5 \cdot 10^{15}$ cm^{-3} (the oxygen concentration does not exceed this value), $q = 0.9$ K/sec, $E = 2.55$ eV and $D(T_m) = 5 \cdot 10^{-9}$ cm^2/sec, we find $M < 10^5$.

Based on the low-angle scattering intensity we estimated the combination

$$CM_1^2 \sim 3 \cdot 10^{20} \text{cm}^{-3}, \tag{24}$$

where M_1 is the number of carriers in a single cloud; C is the cloud concentration in the crystal.

Since $M_1 \leqslant M$, $CM \geqslant 3 \cdot 10^{20}$ cm^{-3}. Based on this inequality and using the estimate obtained above for M, we find $C \geqslant 3 \cdot 10^{10}$ cm^{-3}. The total number of carriers in all clouds (per unit of volume of the crystal) is equal to M_1C; multiplying (24) by C we find $M_1C \geqslant 3 \cdot 10^{15}$ cm^{-3}. With the estimated concentration of clouds C there is a strong cloud overlap, since their carriers (with a total concentration of $3 \cdot 10^{15}$ cm^{-3}) are nearly uniformly distributed in the crystal. Moreover, the measured hole concentration amounts to only $2 \cdot 10^{12}$ cm^{-3}. Consequently this model results in a contradiction and must be rejected.

2. *Dissolution of oxide inclusions.* In this model of cloud formation the key stage is the dissolution of inclusions at $T_f < T_m$. This conclusion was based on formula (21) and the D(T) relation found from experiments on the increase in cloud size a during isothermic annealing (the quantity D (T) is in agreement with the known diffusion coefficient of oxygen in silicon).

The question then occurs: Why are the impurity inclusions not dissolved in the melt (or in the crystal during the cooling process from T_m to T_f)? The answer may be that these impurity particles are not wetted by the silicon, i.e., in both the liquid and solid phases they are separated from the surrounding medium by a narrow gap and therefore the transition of the impurity atoms from the inclusions to the surrounding medium occurs very slowly.

However the situation may change radically in a cooled dislocation-free crystal due to supersaturation by intrinsic interstitial Si_1 atoms. These atoms are absorbed at the matrix boundary and attach to the Si lattice, which compresses the inclusion to a certain pressure p. Further absorption of Si_1 atoms requires completion of work $p\Omega$ on each atom, where Ω is the volume per atom in the Si lattice. The pressure increases until equilibrium is achieved with respect to the absorption reaction described by the equation $p\Omega - \mu_i = 0$, where $\mu_i = kT\ln(C_i/C_{ie})$ is the chemical potential of the interstitial solution (C_{ie} is the equilibrium concentration, while C is the actual concentration of Si_I atoms). Therefore in a supersaturated interstitial solution ($C_i > C_{ie}$) the inclusions are compressed to the equilibrium pressure

$$P = \frac{kT}{\Omega}\ln(C_i/C_{ie}), \tag{25}$$

which, for example, when $C_i/C_{ie} \sim 10$ is equal to 25 kilobars. It is natural to assume that at a sufficiently high p (i.e., with a sufficiently large difference $T_m - T$) the impurity inclusion is pressed "close" to the matrix, combines with the matrix and rapid dissolution of the inclusion then occurs.

It is assumed in this inclusion dissolution mechanism that at T_f, Si_I atoms predominate in the crystal rather than vacancies. In this case if the dominant defect takes the form of vacancies then naturally there will be no supersaturation of the interstitial solution. Therefore we may expect that in this case the rate of dissolution of the inclusions remains small (due to the gap between the inclusions and the matrix) and diminishes with a reduction in T. Qualitatively we may assume that the majority of the dissolved impurity forms at T_m, and that the dimensions of the impurity clouds are determined by formula (21) when $T_f = T_m$.

In both cases the impurity profile around the inclusion will be described by expressions (20), (21). The minimum inclusion size that may be determined proceeding from the assumption of an oxygen concentration in the cluster of $\sim 10^{16}$ cm^{-3} with the inclusion consisting entirely of oxygen was of the order 0.1 μm.

The formation mechanism of carbon clouds is evidently analogous to the formation of oxygen clouds and involves the dissolution of carbide inclusions. Naturally we cannot exclude the possibility of the existence in the silicon of even smaller impurity clouds (such as phosphor clouds).

The proposed formation mechanism of oxygen clouds was tested together with the validity of formula (21) by investigating the influence of a change in the growth rate of the crystal on the impurity cloud size in Si.

In the present study we investigated two crystals grown in identical conditions with a variable growth rate v_p: One crystal was grown from 4 to 15 cm/hr and the second was grown from 15 to 45 cm/hr. The value of the axial temperature gradient G_0 in the 1400-1100°C range (this temperature interval was the precise interval of interest from the viewpoint of cloud formation) was carefully measured [51] and was of the order 300 K/cm.

Varying the growth rate over a broad range and employing exact measurements of the axial temperature gradient make it possible to correctly formulate an experiment to test formula (21).

Figure 20 shows the experimental dependence of the oxygen and carbon cloud size on v_p. Up through $v_p \sim 30$ cm/hr it is clearly described by formula (21) assuming that the size of the oxygen clouds at $v_p \sim 25$ cm/hr is equal to 5.6 μm (experimental size: 5.5-5.6 μm), with the carbon cloud size at 25 cm/hr equal to 2.7 μm. Carbon clouds are observed up to $v_p \sim$ ~ 15 cm/hr. The dimensions of the carbon cloud at the point $v_p = 25$ cm/hr were calculated by formula (21) based on the experimental size of the oxygen cloud (assuming an identical temperature T_f and activation of the oxygen and carbon clouds by identical impurities). Increasing the rate of growth above 30 cm/hr serves to increase oxygen cloud size (from ~ 5 μm to 7-8 μm), which will later manifest a trend towards reduction with growth of v_p.

These results demonstrate that in dislocation-free Si crystals grown with $v_p < 30$ cm/hr cloud formation results from impurity diffusion from the point source. Diffusion is initiated at T ~ 1600 K and the value of T_f in this interval is virtually independent of the growth rate.

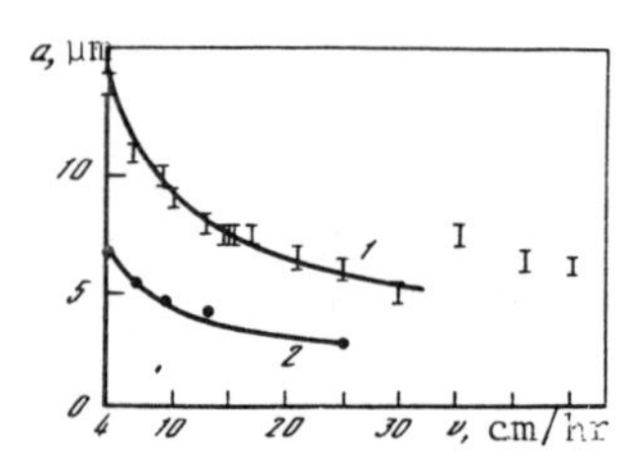

Figure 20. Impurity cloud size a plotted as a function of growth rate.

1 - Oxygen clouds; 2 - carbon clouds.

The sharp increase in oxygen cloud size with an increase in the growth rate above 30 cm/hr is most likely related to the change in temperature T_f. If we estimate T_f by formula (21) for $v_p > 30$ cm/hr, the value of T_f is close to the melting point of silicon (~ 1680 K).

Based on an analysis of the experimental data [52] has proposed a theory of formation of structural defects in dislocation-free Si. This study has noted that in Si crystals at a certain value of v/G there is a reversal of the primary type of intrinsic defects. At $v/G \leqslant 3.3 \cdot 10^{-5}$ cm^2/(sec · K) interstitial atoms dominate, while at $v/G > 3.3 \cdot 10^{-5}$ cm^2/(sec · K) vacancies predominate. The value of v/G at which the cloud size increases (due to the change in v) is $\sim 3 \cdot 10^{-5}$ cm^2/(sec · K), i.e., this is in excellent agreement with the value determined in [52].

Analogous results were obtained from an investigation of one additional silicon crystal whose growth rate varied from 3 to 10 cm/min with an axial gradient of ~ 200 K/cm. The value of v/G at which the jump occurred also was $\sim 3 \cdot 10^{-5}$ cm^2/(sec · K). In addition selective etching was used with this sample to observe the transition from C- to D-defects, corresponding to the defect conversion [52]. This transition occurs in the same region where an increase in oxygen cloud size was observed. Therefore the observed dependence of cloud size on growth rate is accurately described within the scope of the model of impurity microinclusion dissolution proposed above.

We should, however, note that according to this model the cloud scattering intensity in crystals grown in the vacancy mode should be significantly less than the cloud scattering intensity in clouds grown in the interstitial mode. The experimentally observed intensities are comparable.

In concluding this section we note that in the crystals grown with a variable growth rate clouds were observed in both the A-Si regions as well as the B-, C- and D-Si regions. This also concerns our assumption that the impurity clouds are not directly related to the A-defects. It should also be noted that a neutron activation analysis of these samples was not performed.

7. The Nature of Scattering Inhomogeneities in Germanium

Figure 8 gives scattering diagrams of germanium crystals with different defect structures grown by the Czochralski technique. The concentration of electrically-active impurities in these crystals was of the order 10^{12} cm^{-3}. The A-Ge and dislocation samples were from the same monocrystal. The dislocations were introduced to the crystal by means of heat treatments. Their concentration was $\sim 10^3 - 10^4$ cm^{-2}. The concentration of large etch pits in the A-Ge was $\sim 10^6 - 10^7$ cm^{-3}. The B-Ge (with a small etch pit concentration of $\sim 10^9 - 10^{10}$ cm^{-3}) was grown in analogous conditions and had approximately the same impurity composition as the sample described above.

We should note that the scattering diagrams of the germanium and silicon crystals with a variable defect structure were similar. In the germanium the scattering intensity by 6-micron particles was most significant in crystals with a high concentration of large etch pits. In silicon cloud scattering correlates with the concentration of large flat-bottom etch pits. Similar to the diagrams for B- and C-Si, the diagram for B-Ge takes the form of a plateau and the scattering intensity by dislocation crystals is comparatively low.

Therefore we may assume that the optical inhomogeneities discovered in germanium with $a \sim 6$-9 μm have the same origin as analogous inhomogeneities in silicon. High-temperature heat treatments analogous to those performed for germanium were carried out to obtain a final answer regarding the nature of the particles observed in germanium by light scattering.

Figure 21 gives the scattering diagrams of A-Ge before and after annealing at T = 750°C for 2 hours followed by quenching. It is clear that, first, the scattering intensity of the slow section of the diagram dropped by more than an order of magnitude. Unlike silicon no restoration of scattering intensity was observed either as a result of long-term (approximately one year) exposure at room temperature nor as a result of heating at 200°C (for approximately 40 hours). Moreover, unlike silicon the 6-micron particle scattering did not disappear entirely but only dropped significantly in intensity.

Second, the size of the scattering particles determined by the remaining part of the slow section of the diagram, as in the case of silicon, increased with an increase in heat treatment time, following a diffusion law

(Figure 22). The diffusion coefficient determined by formula (22) was equal to $(2.1 \pm 0.5) \cdot 10^{-11}$ cm^2/sec and was in agreement with the familiar diffusion coefficient of oxygen in germanium ($2.0 \cdot 10^{-11}$ cm^2/sec [53] or $2.5 \cdot 10^{-11}$ cm^2/sec [54]).

These results support the conclusion that the nature of the 6-micron inhomogeneities in germanium is identical to that of the impurity clusters in silicon. Evidently they consist of oxygen clouds where their component oxygen, as in the case of silicon, generated electrically-active complexes with rapidly-diffusing impurities. As a result of the heat treatments the complexes dissociated and evidently the majority of rapidly-diffusing impurities escaped from the cloud. The residual scattering remaining after the heat treatments could be either directly attributed to oxygen clouds (without activation) or to residual activation (due to the incomplete dissociation of the complexes at 750°C).

The lack of recovery after heat treatment may be attributed to the fact that the rapidly diffusing impurities are bound in the crystal bulk (either precipitating entirely in the quenching process or bound together in complexes outside the clouds). The second possibility is related to the appearance of thermal donors and thermal acceptors in the germanium resulting from similar heat treatments [44]. Approximately 10^{14} cm^{-3} thermal acceptors were recorded in the test samples by means of the Hall effect. The total concentration of the activating impurity in the clouds was $\geqslant 5 \cdot 10^{13}$ cm^{-3}.

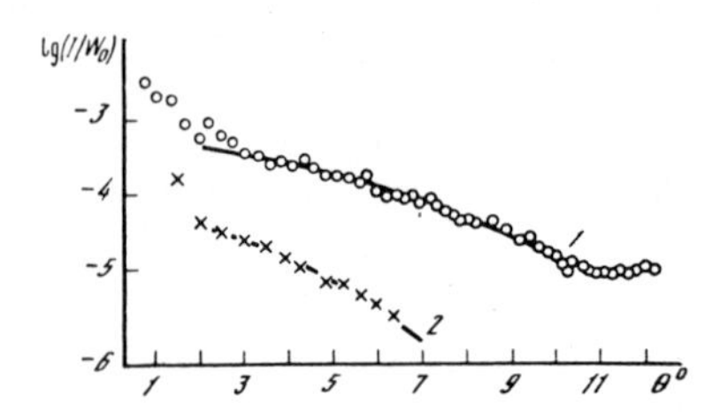

Figure 21. Scattering diagrams of A-Ge.

1 - Original sample; 2 - heat treated sample at T = 750°C for 2 hours.

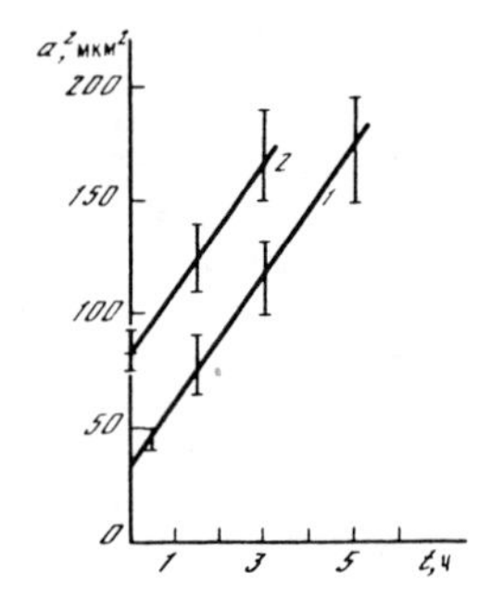

Figure 22. Dimensions of the scattering inhomogeneities *a* plotted as a function of annealing time at T = 750°C.

1 - A-Ge; 2 - B-Ge.

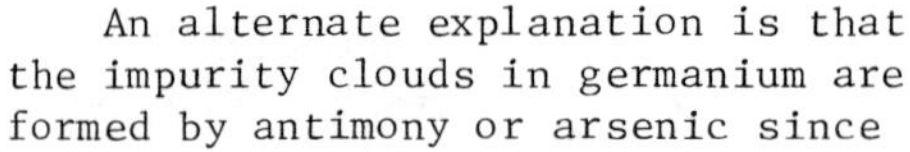

An alternate explanation is that the impurity clouds in germanium are formed by antimony or arsenic since their diffusion coefficient in germanium is close to the diffusion coefficient of oxygen. However the diffuse spread of such clouds would serve only to increase their size and clearly will not be accompanied by a sharp drop in scattering intensity. This explanation is also less likely due to the noticeable differential in the measured diffusion coefficient D* and the value of D for antimony and arsenic (moreover due to the total ionization of these impurities $b = 1$ and D* will coincide with D).

It would be possible to propose that the scattering in the original and heat-treated crystals was attributable to different clouds, however the original size a_0 for the residual clouds (found by extrapolating the curve in Figure 22) coincides with a_0 for the original clouds which indicates an identical nature of the clouds before and after heat treatments. It is also significant that oxygen is always present in the impurity microinclusions in the germanium at the same time that antimony and arsenic are not detected.

Therefore we may say with some confidence that the nature of the 6-micron particles is identical in germanium and silicon: Oxygen clouds activated by rapidly diffusing impurities.

The influence of various heat treatment conditions on B-Ge crystals is of interest. It turned out that the scattering diagrams of B-Ge vary significantly at temperatures of $T \sim 200°C$ (Figure 23). The diagram of the original crystal takes the form of a high-intensity plateau. Scattering by inhomogeneities with $a \sim 8$ μm is observed after heat treatments. Heat treatment experiments at 750°C have revealed that the nature of the inhomogeneities appearing in these crystals is analogous to the scattering particles in A-Ge (Figures 22, 23).

Evidently phase transformations occur in the small inhomogeneities whose scattering produces the plateau in the diagram; these transformations result in a transition to a weakly-scattering form. It is difficult to identify the nature of this process, since we cannot identify the inhomogeneities responsible for the appearance of the small etch pits (evidently scattering by these pits produces the plateau). Heat treatments at 200°C cause the scattering intensity of the plateau to drop sharply which made it possible to observe scattering by the oxygen clouds whose intensity in the original crystals was less than the scattering intensity by small inhomogeneities. It is possible that additional activation of the clouds occurs as a result of the breakdown of the small inhomogeneities (analogous to silicon).

The difference in the dimensions a_0 for the various germanium crystals may be attributed to the changes in T_f.

We will now consider one interesting experiment that made it possible to obtain additional information on the nature of the formation of the impurity clouds. As noted above the dissolution of the inclusions does not begin immediately after their trapping by the oxygen, but at a lower temperature. This fundamental conclusion requires, in our view, direct experimental verification.

We may carry out the necessary test by growing the crystal in a special condition: Stop-growth where at a certain point crystal extrusion from the melt ceases and is excited over time period τ at the previous rate v.

During the growth stop the initial temperature distribution T (z) along the crystal length as shown schematically in Figure 24, a, is conserved in the crystal in the first approximation. If the conclusion discussed above is valid, then during the growth stop process the crystal section of length z_f adjacent to the crystal-melt boundary having a temperature above T_f will no longer contain dissolved inclusions. These inclusions will dissolve only after growth continues; therefore the final dimensions a of the impurity clouds in this section of the crystal will have the same value of a_0 as in a crystal grown in normal conditions without the stop-growth mode. On the other hand the remaining section of the crystal ($z > z_f$) will contain the impurity clouds whose dimensions a will increase over time in accordance with general formula (22) due to the diffusion of the impurity. Consequently the final dimensions of the impurity clouds $a(z)$ in this crystal region are determined by the formula

$$a^2(z)=a_0^2+4D(T)\tau, \tag{26}$$

where T is the temperature at the given point z during the stop in growth. The increment in size a caused by the growth stop will diminish rapidly with z due to the reduction in temperature T and the corresponding rapid drop in the diffusion coefficient D(T). Therefore the change in the dimensions of the impurity clouds a horizontally through the crystal will take the form shown in Figure 24, b.

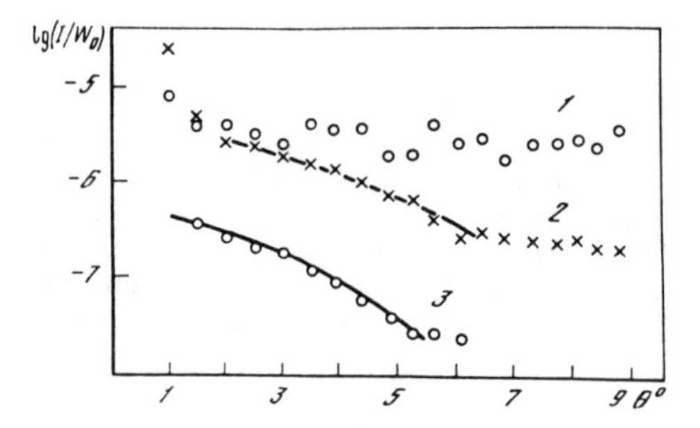

Figure 23. Scattering diagrams of B-Ge.

1 - Original sample; 2 - heat treated sample at T = 200°C for 2 hours; 3 - after heat treatments at T = 750°C (1 hour).

This study has investigated two crystals of dislocation-free germanium (v = 1.6 mm/min, G = 50 K/cm) grown using the stop-growth technique for 10 and 30 minutes. Wafers 2 mm thick were cut along the crystal axis. The dimensions of the impurity clouds $a(z)$ were measured using the technique described above by low-angle scattering of CO_2-laser emission at room temperature. The probe beam (2-3 mm in diameter) was scanned along the Ge wafer in 3 mm increments.

Figure 25 gives the $a(z)$ experimental curves; the distance horizontally along the crystal z was measured from the crystal-melt boundary during the growth stop. The nature of the $a(z)$ relation is in good agreement with the expected relation (Figure 24, b), i.e., it confirms that the impurity clouds are formed at a temperature T_f significantly below the crystallization temperature T_m. We may find T_f by the position of the boundary z_f

between the crystal section with the unchanged impurity clouds ($a = a_0$) and the crystal section with an increase in size a. The temperature gradient G in the high-temperature section of the growing crystal may be taken as constant, i.e., we may assume a linear temperature distribution:

$$T(z) = T_0 - Gz. \quad (27)$$

In this case $T_f = T_0 - Gz_f$. Unfortunately the value of z_f shown in Figure 25 has significant ambiguity since the crystals were cut into several segments and there are no experimental points near these cuts (since the probe beam was not able to get closer than 4-6 mm to the edge of the sample). Hence the value of T_f has significant uncertainty: $T_0 - T_f =$ = 35-85 K (stop time of 10 minutes), and $T_0 - T_f$ = 35-75 K (stop time of 30 minutes).

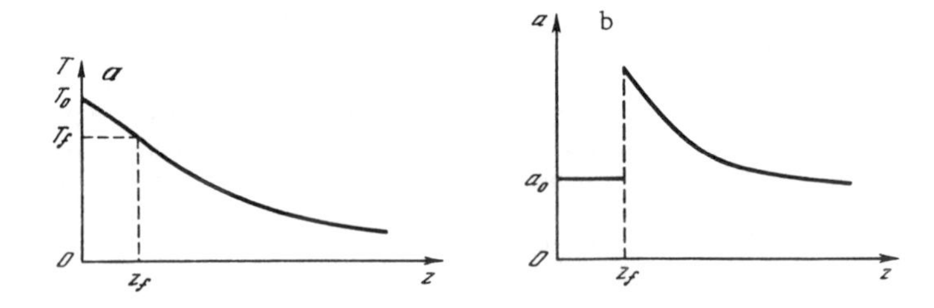

Figure 24. Temperature distribution lengthwise along the growing crystal (a) and the corresponding dimensional distribution a of the impurity clouds expected for the case where a crystal is grown using the stop-growth technique (b).

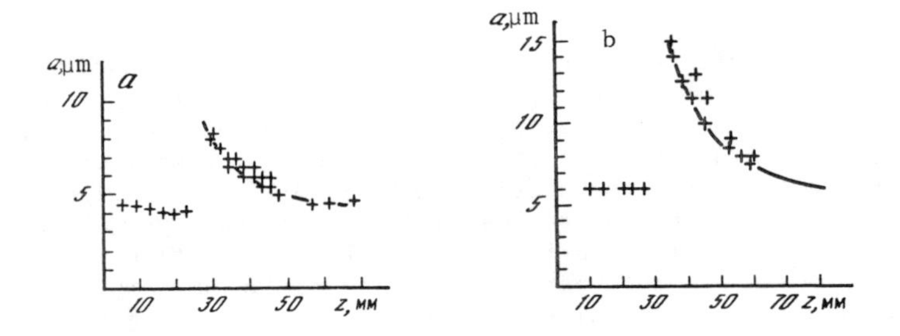

Figure 25. Change in impurity cloud size lengthwise along two germanium crystals grown using the stop-growth technique.

a - Stop time: 10 minutes; b - stop time: 30 minutes.

Using the known temperature dependence of the diffusion coefficient ($cm^2 \cdot sec^{-1}$) of oxygen in germanium

$$D(T) = 0{,}185\exp(-2/kT) \qquad (28)$$

and temperature profile (27) we may draft a theoretical curve for $a(z)$ using formula (26) (the solid line curve in Figure 25). The good agreement with experimental data is additional confirmation of the model of impurity cloud formation.

Therefore we may state with confidence that in both germanium and silicon the slow section of the diagram reflects scattering by oxygen clouds activated by rapidly diffusing impurities formed as a result of the dissolution of impurity inclusions.

The detected impurity clouds represent a new type of impurity cluster observed for the first time; such impurity defects may be formed in any melt-grown crystals. We have observed oxygen clouds in virtually all the test crystals (both dislocation and dislocation-free silicon crystals) grown by the float-zone technique as well as germanium crystals (with the exception of those grown in hydrogen media). The features observed for the various types of crystals will be discussed in the next section.

In conclusion we note that the results from this chapter confirm our assumptions in the calculation formula.

Indeed, as indicated by the established nature of the clouds the deviation in the dielectric constant is small within these clouds. Likewise the variance of the clouds in the crystal sections grown in identical conditions is also small. The validity of the cloud model is confirmed by all experiments described above which, in our view, cannot be interpretated on the basis of any other hypothesis.

8. Impurity Centers in Pure Silicon and Germanium Crystals Grown in Different Atmospheres

The preceding sections of this chapter have focused primarily on results from an investigation of vacuum-grown pure dislocation-free germanium crystals and group I p-type silicon crystals. As noted above the primary inhomogeneities recorded by means of CO_2-laser emission scattering in this material were large-scale non-Gaussian inhomogeneities, Gaussian particles (impurity clouds) and minor particles, and the dominant type of defect in the original crystals were impurity clouds. In this section we will examine results from studies of other pure silicon and germanium samples: n-type silicon crystals, ultra-pure silicon crystals, silicon and germanium samples grown in hydrogen and argon-hydrogen atmospheres, dislocation crystals, etc.

We will first examine the influence of the conductivity type of the sample on scattering in group I A-Si. Figure 26 shows scattering diagrams of p- and n-type crystals characteristic of this group of samples; these samples were cleaved from the same monocrystal. The thermal growing conditions of the crystals as well as the composition of impurities determined by means of neutron activation analysis were similar. The samples were differentiated only in a difference concentration of minor donors and acceptors representing only a weak background with respect to the total concentration of impurities.

It was determined that the *p*-type samples were characterized by oxygen clouds with $a \sim$ 6-7 μm. In *n*-type crystals either Gaussian inhomogeneities with $a \sim$ 8-9 μm were observed or non-Gaussian inhomogeneities with $a \sim$ ~ 10-11 μm were found; we should note that in one of the test A-Si crystals the situation was reversed: Oxygen clouds with $a \sim$ 6-7 μm were observed in the *n*-type samples while larger clouds and non-Gaussian particles with $a \sim$ 7 μm and $a \sim$ 10-11 μm, respectively, were observed in the *p*-type Si.

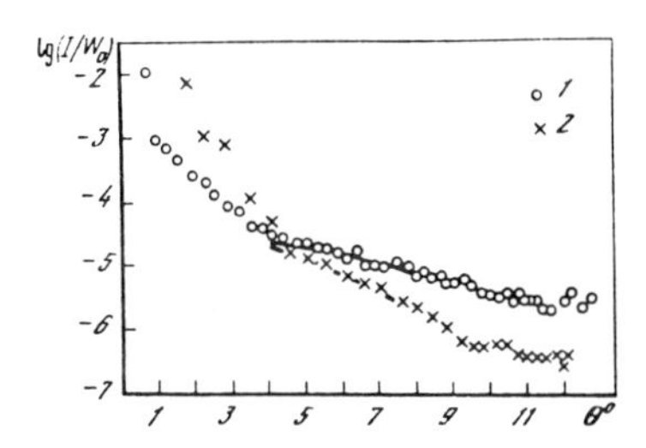

Figure 26. Scattering diagram of group I A-Si.

1 - *p*-type Si; 2 - *n*-type Si.

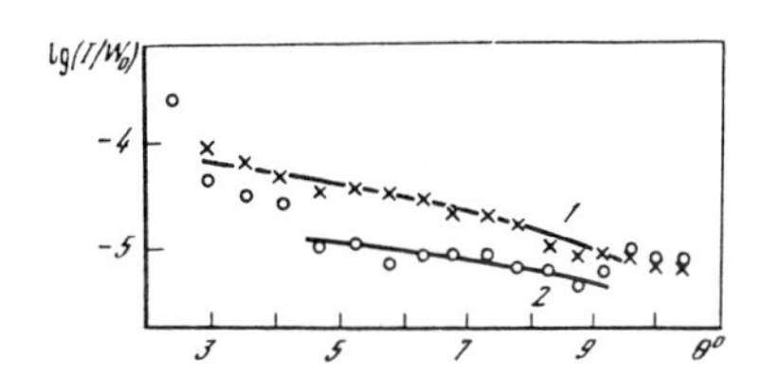

Figure 27. Scattering diagram of dislocation silicon.

1 - *p*-type; 2 - *n*-type.

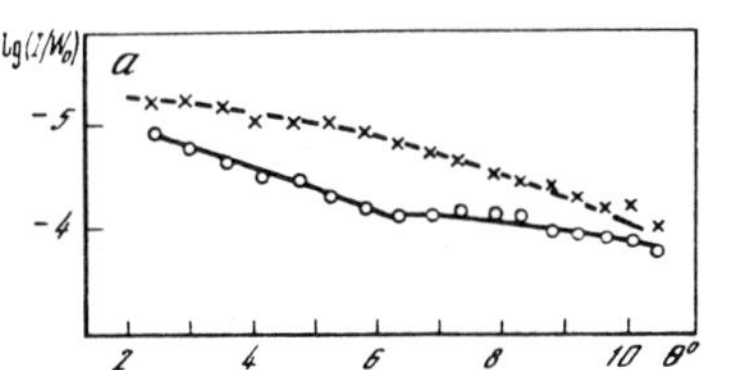

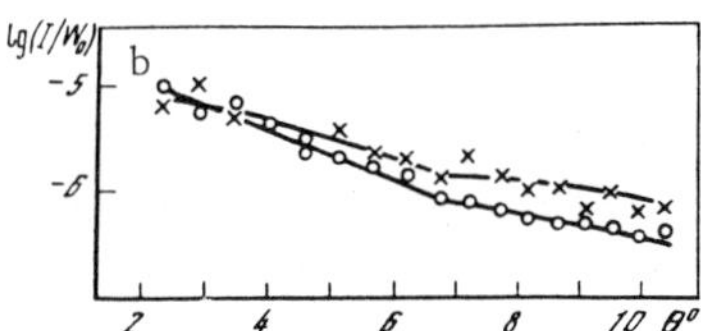

Figure 28. Typical scattering diagrams of group II dislocation-free silicon.

a - Swirl silicon; b - swirl-free silicon.

In the majority of cases the experiments described above resulted from the investigation of crystal sections more than 5 mm from the crystal edge. Investigations of the crystal regions in direct proximity to the sample edges has revealed that the cloud size is somewhat reduced and, moreover, non-Gaussian inhomogeneities with $a \sim$ 7-10 μm are also often observed. The reduction in size of the impurity clouds in the crystal regions adjacent to the crystal edge may be attributed to the somewhat higher cooling rate compared to the crystal center as well as the reduction in T_f (related to the escape of interstitial atoms to the surface).

Figure 27 provides standard scattering diagrams of *p*- and *n*-type pure silicon with a growth dislocation density of ~ 10^3 cm^{-2}. It is clear that as in the case of A-Si crystals, the dominant defect in the *p*-type samples are oxygen clouds, while in *n*-type samples the dominant defect takes the form of non-Gaussian inhomogeneities.

The study also investigated ultra-pure silicon whose impurity concentration (determined by means of neutron activation analysis) was significantly less than in the group I silicon (see Table 2). Standard scattering diagrams of such crystals are shown in Figure 28, a, b. It is clear that the scattering intensity by such samples is significantly (greater than an order of magnitude) less than scattering by group I crystals.

The following may be noted: The A-Si crystals in this group are characterized by impurity clouds with a ~ 6-8 μm. Moreover, non-Gaussian inhomogeneities with a ~ 9-10 μm have also been observed in these crystals. The predominant observed defect in swirl-free silicon crystals were non-Gaussian inhomogeneities with a ~ 7 and a ~ 9-10 μm. Oxygen clouds with a ~ 4-5 μm (see Figure 28, b) were also recorded. It was also determined that heat treatments at 1150°C for several hours have virtually no influence on the non-Gaussian inhomogeneities.

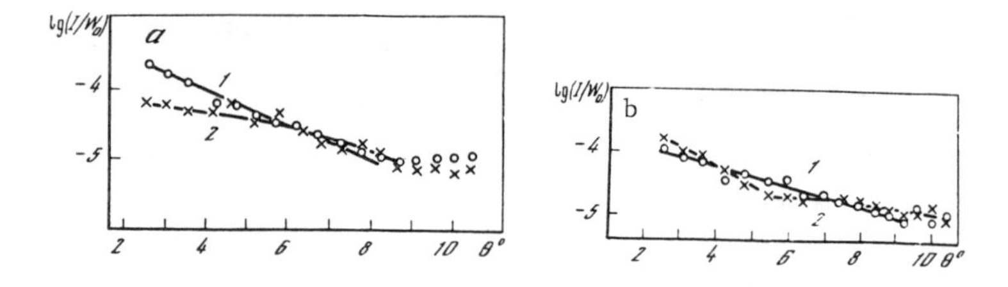

Figure 29. Scattering diagrams of *p*-type and *n*-type A-Si.

a - 1 - *p*-type A-Si before heating; 2 - after heating at 150°C for 1 hour; b - 1 - *n*-type A-Si before heating; 2 - after heating.

We will now examine the features characteristic of dislocation-free silicon grown in argon. The test results for this material were analogous to results from investigations of pure silicon grown in vacuum. The primary observed defect in A-Si crystals were oxygen clouds. Both oxygen clouds and non-Gaussian inhomogeneities with a ~ 16 μm and a ~ 9-10 μm were observed in B- and C-Si crystals. We note that the dimensions of the oxygen clouds in crystals grown in an argon atmosphere are normally somewhat greater than in vacuum-grown samples. Thus, the dimensions of the clouds in A-Si were ~ 10 μm while in B-Si they were ~ 8 μm. In addition we note that the n- and p-type samples were not identical: The n-type Si is characterized by oxygen clouds while non-Gaussian and minor particles were observed in the p-type Si.

The statement that the optical Gaussian inhomogeneities observed in Si crystals grown in an argon atmosphere are oxygen clouds was verified by means of the isothermal heat treatments described in § 5.

We will now consider results from an investigation of the influence of low-temperature (below 200°C) heat treatments on light scattering by pure silicon samples. These experiments have normally been carried out on an assembly for investigating the influence of sample temperature between 300 and 500 K on light scattering (see Chapter 1).

Figure 29 shows the scattering diagrams of p- and n-type A-Si crystals before and after heating to T = 150°C for 1 hour. It is clear that in the case of n-type crystals heating produced particles with $a \sim 9$ μm. However after approximately 20 days at room temperature the initial scattering diagram was almost totally recovered. In p-type silicon heating would result in the vanishing of non-Gaussian particles with $a \sim 7$ μm. Restoration of the scattering diagram by exposure to room temperature occurred in less than 24 hours. When after the heating process the sample was submerged in liquid nitrogen, scattering diagram restoration was not observed for less than 3-4 days. Such lack of thermal stability was characteristic of the majority of Si original samples grown in vacuum and an argon atmosphere which is a serious hindrance in investigating the influence of sample temperature on light scattering in the temperature range above 300 K.

We should note that the pure silicon samples subjected to high-temperature ($\geqslant$ 1000°C) heat treatments followed by slow cooling or additional heating (for example to 600°C) were thermally stable with respect to low-temperature heat treatments (at a minimum up to 250°C). At the same time crystals subjected to high-temperature heat treatments followed by quenching were not thermally stable with respect to low-temperature heating, as in the case of the original material.

The following inhomogeneities were recorded in pure silicon crystals by means of the small-angle light scattering technique: Gaussian particles (impurity clouds) whose dimensions were dependent on the growing conditions, non-Gaussian particles with $a \sim$ 18-20 μm; $a \sim$ 9-10 μm and $a \sim 7$ μm as well as "minor" particles. The nature of the latter inhomogeneities is not known. Since the non-Gaussian defects are not temperature stable in a number of cases even when heated to 100-200°C, we may assume that they also are weak clusters of non-temperature-stable, electrically-active complexes.

We will now proceed to a description of the features characteristic of pure, dislocation-free crystals grown in hydrogen and argon-hydrogen atmospheres. Figure 30 gives scattering diagrams characteristic of these crystals. Pure, dislocation-free silicon grown in a hydrogen atmosphere normally contains Gaussian inhomogeneities with $a \sim$ 12-14 μm and $a \sim$ 4-5 μm.

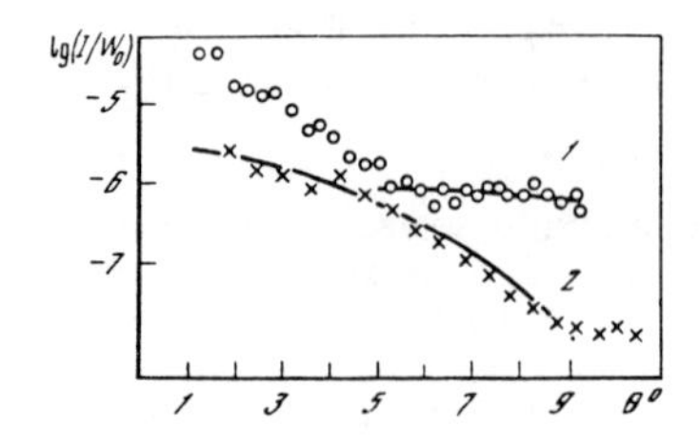

Figure 30. Scattering diagram of silicon grown in hydrogen.

1 - 100% H_2; 2 - Ar_2 + 2% H_2.

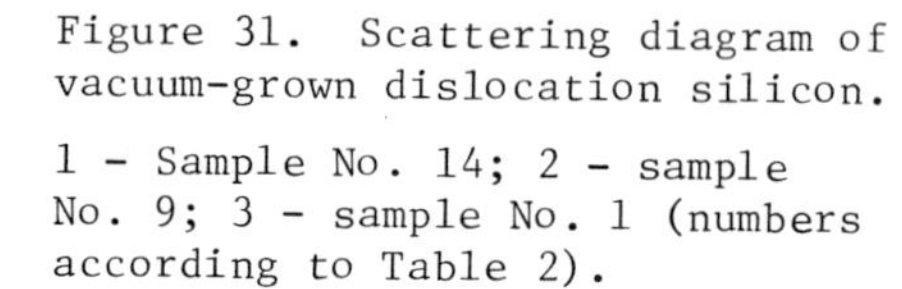

Figure 31. Scattering diagram of vacuum-grown dislocation silicon.

1 - Sample No. 14; 2 - sample No. 9; 3 - sample No. 1 (numbers according to Table 2).

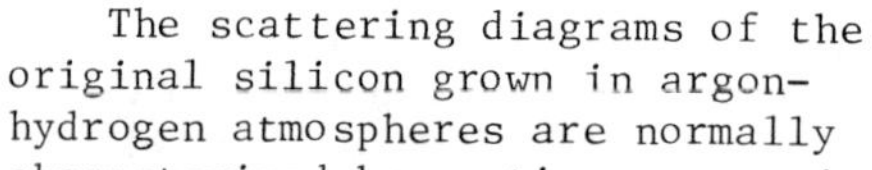

The scattering diagrams of the original silicon grown in argon-hydrogen atmospheres are normally characterized by sections approximated by scattering by large (with $a >$ > 25 μm) inhomogeneities. Particles with $a \sim 8$-10 μm and minor particles have also been observed. Occasionally silicon samples grown in argon-hydrogen atmospheres are characterized by weak scattering by minor particles (it is this material that is used to fabricate cryostat windows), although heat treatments at 400°C will produce large inhomogeneities in these crystals. The nature of the optical inhomogeneities observed in silicon grown in an atmosphere containing hydrogen is not known at the present time.

We will now consider results from an investigation of ultra-pure germanium with a growth dislocation density of $\sim 10^3 - 10^4$ cm^{-2} grown both in vacuum and in hydrogen atmospheres. The interest in this material has resulted from its broad utilization in manufacturing λ-emission detectors. This study employed multiple samples of ultra-pure germanium grown at various facilities. The difference donor and acceptor concentration ($|N_Д - N_A|$) in the test crystals was varied over a broad range: From $3 \cdot 10^{10}$ cm^{-3} to $5 \cdot 10^{12}$ cm^{-3}.

We will first examine the results from an investigation of vacuum-grown ultra-pure germanium. Characteristic scattering diagrams for such crystals are shown in Figure 31. We may identify the following typical defects: Large non-Gaussian inhomogeneities ($a \sim 18$-20 μm); Gaussian inhomogeneities with dimensions varying from 5 to 9 μm and minor particles. Intense scattering by large ($a > 25$ μm) inhomogeneities was occasionally observed in a number of samples with anomalous electrophysical properties (see Chapter 3). The dislocation germanium scattering intensity varied over a significant range (~ 3 orders of magnitude).

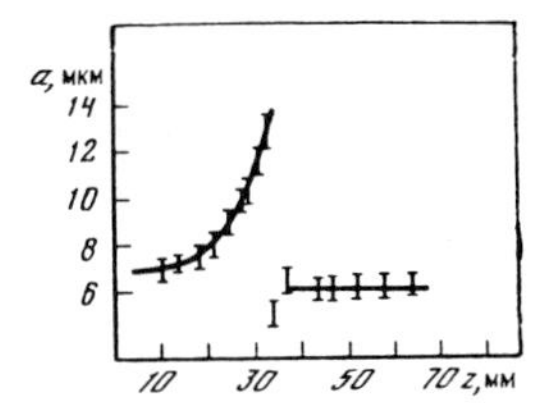

Figure 32. Change in dimensions of impurity clouds lengthwise along dislocation silicon crystal.

Stop time: 15 minutes.

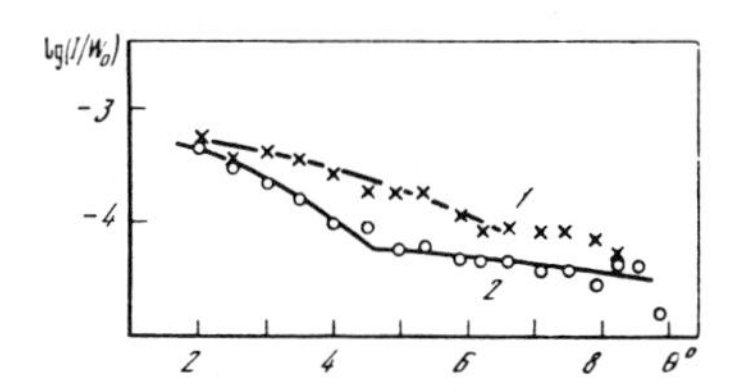

Figure 33. Scattering diagrams of vacuum-grown dislocation silicon.

1 - Original crystal; 2 - crystal heated to T = 150°C for 2 hours.

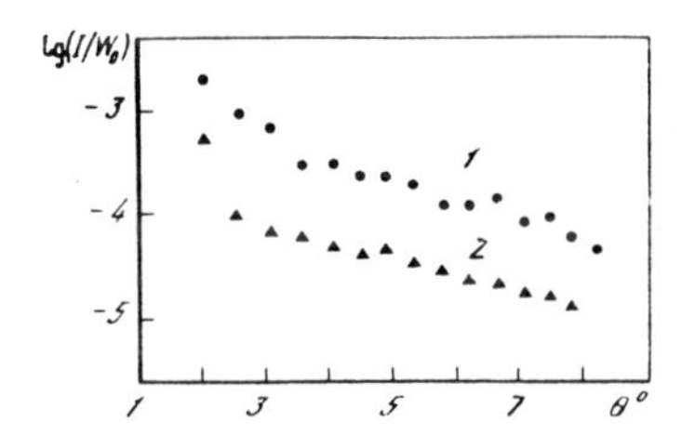

Figure 34. Scattering diagram of hydrogen-grown dislocation germanium.

1 - T = 300 K; 2 - T ~ 150 K.

We assumed that the Gaussian particles, as in the case of dislocation-free germanium, are oxygen clouds. We confirmed this assumption by the experiment described in § 6 devoted to the influence of stop-growth of the crystal on the dimensions of the scattering inhomogeneities in the germanium. A sample with ($|N_Д - N_A|$) ~ 10^{12} cm^{-3} and a dislocation density of ~ 10^3 cm^{-2} was employed. Gaussian particle scattering dominated the scattering diagrams of this sample. The dependence of the dimensions of these particles on the distance to the growth stop point are given in Figure 32 and are accurately described assuming that they are oxygen clouds that formed at a temperature T_f ~ 1150-1160 K.

The investigation of the influence of low-temperature heating has demonstrated that the impurity clouds in dislocation silicon, as in the case of pure silicon, includes non-temperature-stable complexes. Figure 33 gives scattering diagrams of the original dislocation germanium and dislocation germanium heated to 150°C. The original sample is characterized by scattering by clouds with a ~ 9-10 μm. After heating scattering by particles with a ~ 7 μm predominates in the diagram. The original diagram is restored after several days of exposure at room temperature.

We also investigated multiple samples of ultra-pure germanium grown in hydrogen at various facilities (including facilities abroad). The scattering diagrams of these crystals were similar in the majority of cases and were characterized by non-Gaussian inhomogeneities with $a \sim$ 7-8 μm. The minimum scattering intensity by these inhomogeneities was in a sample with $(|N_Д - N_A|) \sim 3 \cdot 10^{10}$ cm^{-3} $(I \sim 10^{-6}$ sr$^{-1})$, with a maximum intensity of $(I \sim 5 \cdot 10^{-4}$ sr$^{-1})$ in a sample with $(|N_Д - N_A|) \sim 5 \cdot 10^{12}$ cm^{-3}.

Figure 34 gives the scattering diagram of one of the hydrogen germanium crystals at T = 300 and 150 K. It is clear that the scattering intensity drops as the sample is cooled. This result indicates that the particles observed in the hydrogen germanium are clusters of electrically-active centers with an ionization energy $\Delta E < 0.15$ eV. The formation mechanism of these clusters is not known at present: The impurity clouds are particles with a different profile distribution function of the dielectric constant

$$f(r,a) = e^{-r^2/a^2} . \tag{29}$$

In addition we investigated germanium samples grown at different growth rates: 0.3 and 2.2 mm/min. The dimensions of the impurity clusters remain virtually the same in these cases at the same time that, as demonstrated in § 5, the dimensions of the impurity clouds are proportional to $1/\sqrt{v}$.

Therefore the results from this section indicate that pure silicon and germanium crystals contain, in addition to impurity clouds, a wide range of different impurity clusters including non-temperature-stable complexes and slowly- and rapidly-diffusing impurities. Evidently "activating" or compensating rapidly-diffusing impurities are exchanged between these clusters.

CHAPTER 3

THE INFLUENCE OF OXYGEN CLOUDS ON THE PROPERTIES OF ULTRA-PURE GERMANIUM

The investigation of impurity clouds is of interest primarily in connection with their possible influence on the properties of semiconductor crystals and the parameters of instruments manufactured using such crystals.

In this chapter we will consider the results from initial experiments in this field.

9. The Influence of Photoexcitation on Light Scattering by Germanium Crystals

Since we assume that light scattering by clouds is caused by changes in the free carrier concentration, the interaction between the nonequilibrium carriers and the clouds is of interest. In the present study we will investigate the influence of volumetric photoexcitation on light scattering by germanium crystals. Volumetric photoexcitation is convenient for such experiments, since in this case it is not necessary to apply contacts or employ chemical etching of surfaces in order to reduce the rate of surface recombination. Moreover, volumetric photoexcitation makes it possible to excite relatively significant areas in the crystal bulk. The experimental set-up to investigate the influence of volumetric photoexcitation on light scattering is described in Chapter 1.

Figure 35 shows the absorption signals of CO_2-laser emission for the case of absorption by the generated carriers together with the scattered light signals for carrier excitation by a $CaF_2:Dy^{2+}$ laser ($\theta \sim 3°$). It is clear that the absorption and scattering pulses are related, i.e., the recorded scattering pulse is caused by changes in the free carrier concentration.

The scattered light pulse waveform is heavily distorted. This may occur, first, due to the use of a low-frequency filter in the network and, second, undetectable, minor changes in the phase shift between the signal ray and the reference ray may influence the pulse waveform. These changes may occur due to, for example, vibrations. Moreover, additional difficulties will also arise in connection to the oscillograph signal of $\sim \sqrt{I_{pac}}$.

The signal value was measured either at its maximum (by setting a short duration strobe to the signal maximum) or by measuring the mean signal level (with the strobe duration in the order of the scattering pulse duration). The measurement results using both techniques were in agreement within acceptable experimental error.

Figure 36 shows the scattering diagrams of a dislocation silicon crystal ($|N_Д - N_A|) < 10^{11}$ cm^{-3}, $N_g \sim 10^3$ cm^{-2}) with and without photoexcitation. It is clear that the scattering diagram with photoexcitation (i.e., the section of the diagram dependent on the free carrier concentration in the sample) consists virtually solely of a single slow section and under photoexcitation its intensity ($n_{HOC} \sim 10^{15}$ cm^{-3}) is two to three orders of magnitude greater than without photoexcitation. Scattering diagrams without photoexcitation have a standard form for dislocation crystals. It follows that the cloud scattering intensity increases with growth of carrier concentration at the same time that evidently the scattering in the fast section remains unchanged.

Figure 37 shows the dependence of I_{pac} on the mean concentration of excited carriers. It is clear that this relation is a square-law relation up through a concentration of $\sim 4 \cdot 10^{15}$ cm^{-3}. Saturation is observed at higher concentrations. The dimensions of the scattering objects evidently increase insignificantly under photoexcitation. Analogous results were obtained from an investigation of dislocation commercial-grade germanium crystals: $n_{пр} \sim 10^{12}$ cm^{-3}.

The influence of temperature on light scattering by germanium crystals was also considered. The experiments were performed at T = 77 and 10 K on a dislocation germanium sample $n_{пр} \sim 10^{12}$ cm^{-3}. Figure 38 gives the scattering diagram for scattering by such a germanium crystal at T = 10 K and a carrier concentration of $\sim 10^{14}$ cm^{-3}. The aperture of the cryostat windows made it impossible to record the scattering diagram at angles of greater than 3°30'. The dependence of I_{pac} on n_{HOC} is nonlinear. The scattering intensity at T = 10 K and at a carrier concentration of 10^{14} cm^{-3} was more than an order of magnitude greater compared to an identical concentration at room temperature.

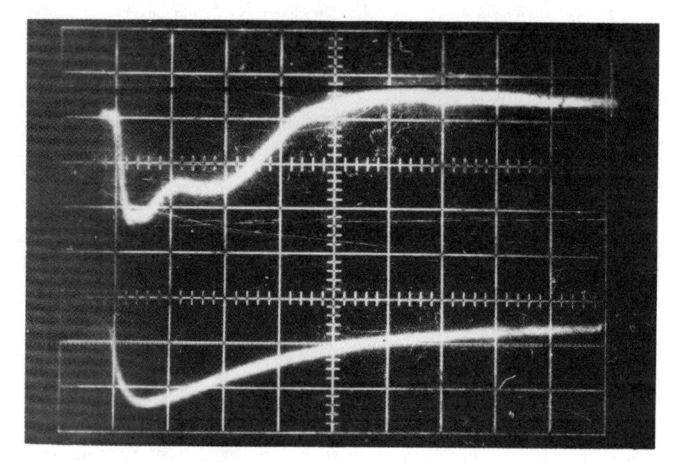

Figure 35. Temporal dependence of a scattered and an absorbed CO_2-laser signal.

Above - scattering signal; below - absorption signal.

Our results are interpreted based on the assumption that the generated carriers are trapped within the impurity cloud. In reality due to trapping the carrier concentration will grow within the cloud at the same time that cloud scattering is caused by the higher carrier concentration and therefore light scattering intensity will also grow. We will examine this process in greater detail on the basis of the following assumptions.

1. The cloud contains a large quantity of a single type of deep trapping centers with a concentration of $N(r)$ (for definiteness we will take these to be donors). In the initial state they are weakly ionized, i.e., $N^+ \ll N$. Such centers may in fact exist, since any impurities that can form complexes with both shallow and deep levels may fall within the cluster.

2. The trap coefficient of the holes at this level j_p is significantly greater than the trap coefficient of the electrons j_n.

For simplicity we will consider the situation where the cluster contains only a single type of impurity with the properties outlined above. Since the dimensions of the cloud a significantly exceed the Debye screening length ($l \sim 1$ μm), the neutrality condition

$$n - p = N^+, \quad (30)$$

will be satisfied, where n and p are the electron and hole concentrations; N^+ is the charged donor concentration.

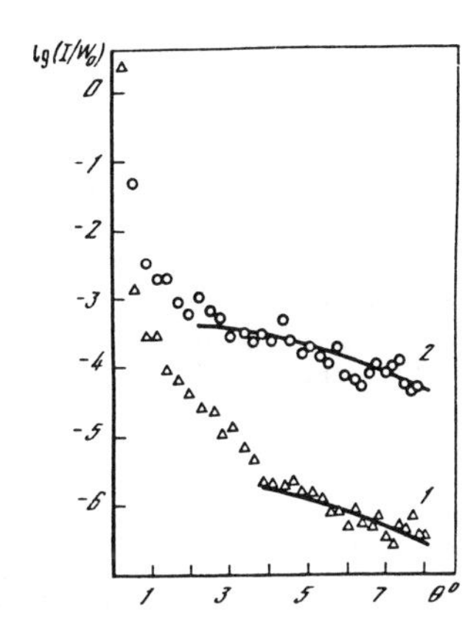

Figure 36. Scattering diagrams of dislocation germanium.

1 - Sample without photoexcitation;
2 - sample under photoexcitation ($n_{HOC} \sim 10^{15}$ cm^{-3}).

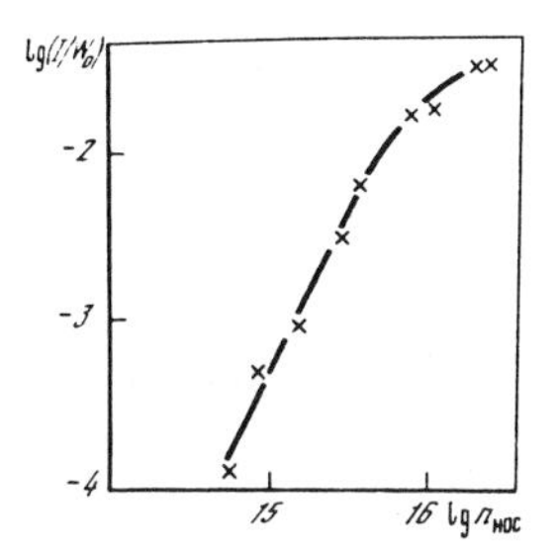

Figure 37. Scattering intensity of dislocation germanium plotted as a function of free carrier concentration.

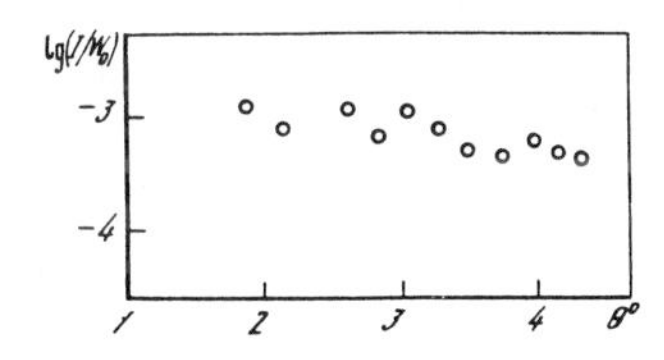

Figure 38. Scattering diagrams of dislocation germanium under photoexcitation ($n_{HOC} \sim 10^{14}$ cm^{-3}) at T = 10 K.

The sharp increase in scattering under photoexcitation (i.e., the majority carrier concentration) is due to the growth of N^+, i.e., the trapping of holes by the neutral donors. The dependence of the stationary concentration N^+ on n and p is given by the familiar expression [55]

$$\frac{N^+}{N} = \frac{j_n n + j_p p_1}{j_n(n + n_1) + j_p(p + p_1)}, \tag{31}$$

where n_1 and p_1 are the reduced state densities of this energy level (for example, $p_1 = N_v g \exp^{-(E-Eg/KT)}$, $E-E_v$ is the distance from the level to the top of the valence band, g is the effective degeneration factor of the level, N_v is the density of states in the valence band).

Since the dimensions a are significantly less than the diffusion length, in the steady-state the electrochemical potentials of the electrons and the holes are constant within the inhomogeneity. Here we have constancy of np:

$$np = n_0 p_0 = n_0^2, \tag{32}$$

where n_0 and p_0 represent carrier concentrations in the semiconductor bulk ($n_0 = p_0$).

As we see from Figure 37 the scattering intensity $I \sim n_0^2$ and the $I \sim n_0^2$ law holds up through concentrations of $4 \cdot 10^{-15}$ cm^{-3}. It follows that in this range of concentrations n is proportional to n_0. Then in accordance with (30), (31) p and N^+ are also proportional to n_0. This is possible if the term $j_p p$ predominates in the numerator, and the term $j_p p_1$ dominates in the denominator, and this condition will be satisfied if $j_p \gg j_n$ (this is one of the conditions imposed on the level) up through concentrations of $\sim 4 \cdot 10^{15}$ the degree of ionization of the level is still small.

In this case (31) is simplified and yields

$$N^+ = pN/p_1. \tag{33}$$

Then we find the desired concentration of majority carriers $n(r)$ in the inhomogeneous region:

$$n(r) = n_0 \sqrt{1 + N(r)/p_1}, \tag{34}$$

with a hole concentration of $p = n_0^2/n$.

Saturation of this relation when $n_0 > 5 \cdot 10^{15}$ cm^{-3} may also be related to the fact that the condition $N^+ \ll N$ is no longer satisfied. Then in the denominator we may ignore the term p compared to p_1 and the relation again becomes saturated. The physical picture here is very simple: When the excitation level is relatively small and $N^+ \ll N$, N^+ will be proportional to n_0. At high excitation concentrations all the levels in the cluster will be ionized and, consequently, a further increase in pumping will not serve to increase the number of ionized levels.

The temperature dependence of the scattering intensity is easily explained within the scope of this model; the scattering intensity will grow since the thermal release of the trapped holes from the impurity levels becomes negligible at T = 10 K. In this regard we evidently cannot, even at relatively low concentrations, ignore p compared to p_1 and I (n_0) will no longer be described by simple formula (33). This is also confirmed by preliminary results from an investigation of I (n_0) at helium temperatures: The scattering intensity I was proportional to $n^{1.7}$.

Now based on this examination and available experimental data we will attempt some preliminary estimates for the electron and hole trapping cross-sections as well as the maximum impurity and cluster concentrations and the depth of the impurity level.

We will assume that, as in the case of zero illumination (when n_0 is equal to the intrinsic concentration of $\sim 10^{13}$ cm^{-3}) the observed scattering is attributable to carriers formed as a result of the ionization of the deep levels. (In principle, of course, the cluster may also contain shallow levels that are nearly completely ionized at room temperature.) The maximum electron concentration in the cluster is $n_m \sim 10^{15}$ cm^{-3}. From here $N_m/p_1 \sim 10^4$, where N_m is the maximum impurity concentration. As noted above it follows from the proportionality of the quantities N^+, p, n (in accordance with (31)) that $p \ll p_1$ which may be written as $n_o \ll \sqrt{N_m p_1}$. This inequality will hold at least to $n_o \sim 10^{15}$ cm^{-3}, and from here $N_m > 10^{17}$ cm^{-3} and $p > 10^{13}$ cm^{-3}.

We should note that the deviation of the I (n_o) relation from a square-law when $n_o > 5 \cdot 10^{15}$ cm^{-3} may not be related solely to the saturation of the deep trapping centers in the cloud. The problem is that at $n_o \sim 10^{16}$ cm^{-3} strong attenuation of the probe beam is initiated due to selective absorption by the holes [36]. Moreover, the very carrier photoexcitation distribution becomes irregular due to the selective absorption of the dysprosium laser emission [35]. Therefore the value $N_m \sim 10^{17}$ cm^{-3} is a lower estimate for the deep level concentration in the cloud.

It follows from our estimate for p_1 that the donor impurity level lies near the center of the bandgap (or even in the lower half of the bandgap). If N_m does not significantly exceed the value of 10^{17} cm^{-3} given above, we may estimate $j_p/j_n \geqslant 10^4$ from the inequality $j_n n < j_p p$, i.e., this level is indeed the trapping center for the holes.

These values may be considered to be estimates only. Indeed we do not know whether or not scattering in the presence of photoexcitation is caused by the ionized deep levels. Unfortunately the lower sensitivity of the assembly for investigating scattering with photoexcitation (see Chapter 1) makes it impossible to perform the experiment at a carrier concentration of $\sim 10^{13}$ cm^{-3} (at room temperature). Extrapolation reveals that, evidently (we should remember that the absolute scattering intensities were determined accurate to a factor of 3) the carrier concentration without photoexcitation in a cluster formed as a result of deep level ionization is 3-4 times smaller than $n_m \sim 10^{15}$ cm^{-3}. Therefore our assumption of a single type of level in the cloud is incorrect, although qualitatively this does not change the discussion above.

Moreover the concentration of large etch pits of dislocation-free origin in the crystals used in the photoexcitation experiments is unknown. The value of $n_m \sim 10^{15}$ cm^{-3} was selected by analogy to A-Ge crystals investigated in § 7. The scattering intensity of the crystals examined in this section was two to three orders of magnitude smaller.

At present it is impossible to exactly determine the origin of the low scattering intensity: Whether it is attributed to the lower concentration of the clouds or their lower activation (i.e., the reduction in n_m). Most likely both factors operate simultaneously. As a result the value of the quantities N_m, j_p/j_n and the level depths will be somewhat lower. For example, if we take a value of $\sim 10^{14}$ cm^{-3} for the case of zero photoexcitation, the estimate of the upper boundary for N_m will be $\sim 10^{16}$ cm^{-3}.

Formula (31) is valid in the case of an equilibrium system, i.e., for stationary carrier excitation. The present study employed pulsed excitation. However since the rise time of the scattered light pulse was less than 10 μsec and the carrier lifetime was greater than 100 μsec we may assume that quasiequilibrium is established in the carrier system. We therefore may, evidently, use formula (31) to examine the regularities of the trapping process in the cloud and for producing approximate estimates.

Another interpretation of the influence of photoexcitation on light scattering by germanium crystals is also possible. It is entirely possible that the carriers are generated in the clouds by the "two-step" generation process through some level (located somewhere in the center of the bandgap). In this case if the concentration of such levels is high ($\sim 10^{17}$ cm^{-3}), the rate of carrier generation due to the two-step process may be higher than in the case of two-photon absorption. Then the situation may be that the carriers in the crystal bulk are excited by two-photon absorption and within the cloud by two-step generation and the rate of generation is significantly higher in the cloud. Therefore the carrier concentration in the cloud may be significantly higher than the average concentration throughout the crystal bulk. Two-step generation in the crystal bulk is low since the concentration of deep levels is low.

A further investigation of light scattering by photoexcitation of crystals is of indisputable interest, primarily focusing on the kinetics of the scattered light pulse, experiments on carrier generation due to single-photon absorption of light with a quantum energy close to the bandgap and detailed measurements at low temperatures. These experiments make it possible to determine the mechanism of photoexcitation influence on carrier concentration in the cloud and to identify the nature of the deep levels as well as to obtain information on the influence of the clouds on the carrier properties.

At present we may identify the following results from investigations of the influence of photoexcitation on light scattering in germanium crystals.

1. It was finally confirmed that cloud scattering in Ge is indeed caused by the existence of a free carrier cluster. We should note that this study was the first to observe light scattering by free carrier clusters in pure silicon and germanium crystals at room temperature.

2. It is established that the clouds in germanium contain a large quantity of deep centers (both models require deep centers) and these clouds may significantly influence the free carrier recombination (generation) processes.

3. Photoexcitation may be used to identify and investigate cloud scattering in germanium crystals even when this scattering is less than scattering by other inhomogeneities. From this viewpoint B-Ge is of special interest.

Figure 39 presents scattering diagrams of the original B-Ge crystal and a B-Ge crystal under photoexcitation. It is clear that oxygen clouds with $a \sim 6$ μm appear under photoexcitation. This result again confirms the fact that oxygen clouds also exist in the original B-Ge crystals.

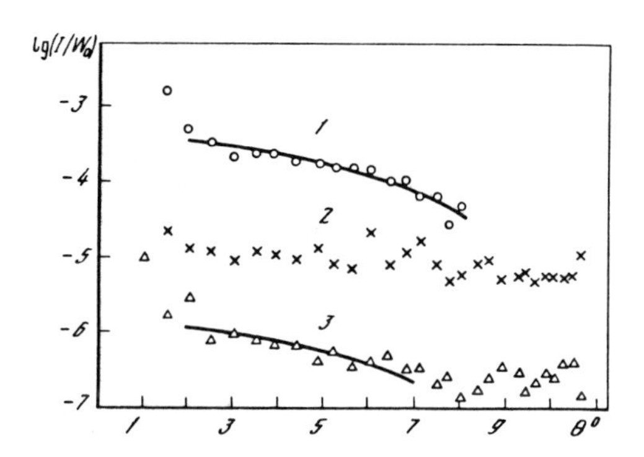

Figure 39. Scattering diagram of B-Ge.

1 - Original sample; 2 - original sample under photoexcitation; 3 - original sample after heating at 350°C.

We note in concluding this section that the present study also observed backscattering (i.e., at 180°) by dislocation germanium under volumetric photoexcitation over a broad temperature range (10-300 K). The backscattering intensity was more than an order of magnitude weaker than the scattering intensity at small angles and also increased with a drop in temperature. The nature of the particles scattering at 180° is still not clear. We should note that the scattering intensity by oxygen clouds in this direction will be negligible (in the order of e^{-500} of the small-angle scattering intensity). Therefore we obviously are observing backscattering by some minor ($a \leqslant 1$ μm) inhomogeneities.

10. The Influence of Impurity Clouds on the Resolution of γ-Emission Detectors Fabricated From Ultra-Pure Germanium

Ultra-pure germanium is widely used today to fabricate γ-emission detectors. As demonstrated by studies performed in the mid 1970s based on the amplitude analysis of calibrated charge transfer in the p^+-n-n^+-structures of nuclear radiation detectors the anomalously large trapping of carriers in pure germanium (and, consequently, the low resolution of the detectors) may be attributed to the potential wells within the scope of the model of large-scale traps [18]. Study [56] assumes that these are weak (with an ionized (at T = 77 K) impurity concentration of $\sim 10^{13}$ cm^{-3}) yet large-scale (~ 10 μm) impurity clusters. It is clear that the parameters of these clusters are close to the parameters of our oxygen clouds.

Table 4

Detector parameters vs. I_{pac}

Sample No.	$W_{ЯM}$, cm^{-3}	Scattering intensity I, W/sr	a, μm	δW, keV	λ*, %
1	$1 \cdot 10^5$	$1 \cdot 10^{-8}$	9	1.2	0.14
2	$1 \cdot 10^5$	$1 \cdot 10^{-8}$	9	1.4	0.19
3	$2 \cdot 10^5$	$1 \cdot 10^{-7}$	8.5	2.5	0.4
4	$4 \cdot 10^5$	$5 \cdot 10^{-8}$	9	3.5	0.5
5	$8 \cdot 10^5$	$1 \cdot 10^{-7}$	9.5	7	1.1
6	$1 \cdot 10^6$	$1 \cdot 10^{-7}$	6	8	1.2
7	$6 \cdot 10^6$	$2 \cdot 10^{-7}$	6.5	8	1.2
8	$1.8 \cdot 10^6$	$5 \cdot 10^{-7}$	6.5	10	2
9	$9 \cdot 10^5$	$2 \cdot 10^{-4}$	6	10	1.5
10	$2 \cdot 10^6$	$5 \cdot 10^{-7}$	6	15	2.5
11	$3 \cdot 10^6$	$1 \cdot 10^{-5}$	6	40	5
12	$6 \cdot 10^6$	$3 \cdot 10^{-6}$	5	43	4.2
13	$1 \cdot 10^7$	$1 \cdot 10^{-6}$	5	43	-
14	$1 \cdot 10^7$	$3 \cdot 10^{-5}$	6.8	43	-

*λ - Relative reduction in accumulated charge.

Accounting for the fact that photoexcitation experiments have revealed that oxygen clouds in germanium may be carrier trapping centers, it becomes clear why their existence in the crystals may cause a significant reduction in γ-detector resolution. Therefore the investigation of clouds and their influence on γ-detector quality is of significant interest.

The present study investigated a significant number of vacuum-grown ultra-pure p- and n-type germanium. Samples with $N_g \sim 10^3 - 10^4$ cm^{-2} made up the majority of germanium samples. A few A-Ge crystals were also investigated. In conjunction with the light scattering experiments, detectors were fabricated and their characteristics were measured. Etching was also used to investigate defect structure of the crystals. Scattering experiments were carried out at room temperature without photoexcitation. Scattering diagrams characteristic of this group of samples are described in § 8. Measurement results are given in Table 4.

An analysis of the results reveals that all good samples ($\eta \sim 3$ keV) yielded weak scattering $< 10^{-7}$ W/sr. The scattering intensity by the crystals with a high value of η was normally an order of magnitude greater and the scattering diagrams of the poor crystals revealed very intense scattering characteristic of clouds with a = 6 μm. The scattering diagrams of the "average" crystals occasionally did not contain a slow section, but rather relatively intense scattering by other inhomogeneities was observed (a plateau and a fast section). Therefore there is direct evidence of a correlation between the scattering intensity in the original material and the quality of instruments fabricated using these crystals.

In our view another interesting result is the correlation discovered in the present study between cloud scattering intensity and the concentration of large etch pits of dislocation-free origin (see Figure 40). This fact confirms our assumption that the oxygen clouds themselves are responsible for the large etch pits. We note that there is no direct proportional relation between scattering intensity and the concentration of the etch pits. This may be attributable to the fact that the scattering intensity is determined not only by the cloud concentration but also by their degree of activation which may differ somewhat in different crystals. Study [57] has established that the concentration of large etch pits correlates with the resolution of the γ-detectors.

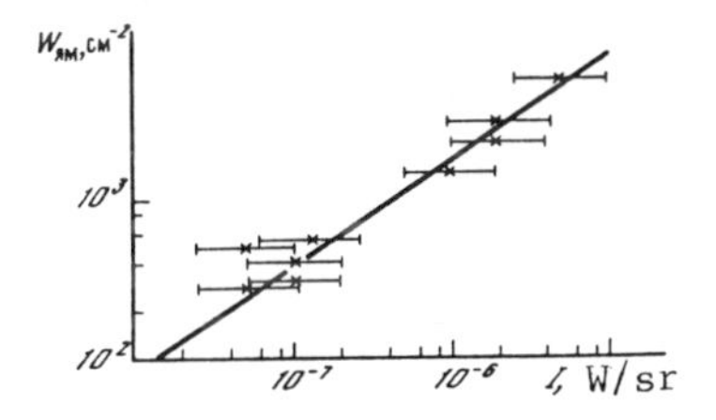

Figure 40. Cloud scattering intensity in ultra-pure germanium plotted as a function of the surface concentration of etch pits.

These data make it possible to propose that the oxygen clouds (manifest as large etch pits) indeed have a significant influence on the quality of γ-detectors fabricated from ultra-pure germanium. Hence we will now consider in somewhat greater detail the possible mechanism of cloud influence on the charge accumulation process in the detector.

Small-angle light scattering by the impurity clouds is caused primarily by a local increase in the free carrier concentration. For definiteness we will assume that the clouds are clusters of acceptor centers. The carrier (hole) distribution is Gaussian which is related to the cloud formation mechanism: The diffusion of oxygen from the initial point sources (oxide inclusions), i.e.

$$p(r) = p_m \exp(-r^2/a^2), \tag{35}$$

where p_m is the maximum hole concentration (at the center of the cloud). The values of a for a number of test samples are given in Table 4 and the values of C are recomputed from the surface pit density, also given in Table 4. The absolute value of the scattering intensity I is proportional to the combination Cp_m^2, i.e., with a known C we may also estimate the maximum local hole concentration at $p_m \sim 10^{15} - 10^{16}$ cm^{-3}. The corresponding Debye screening length (in the order of fractions of a micron) is significantly less than a, i.e., the cloud is quasineutral and the hole distribution $p(r)$ is identical to the distribution of charged acceptors (oxygen complexes with a certain rapidly diffusing impurity). If the acceptors are totally ionized at T_K (room temperature at which scattering is measured), their concentration $N_A(r)$ coincides with $p(r)$. In the opposite case (weak ionization, i.e., a rather deep acceptor level) the value $p = N_A$ is proportional to N_A/p, i.e., $N_A(r)$ is proportional to $p^2(r)$. In both limiting cases the acceptor distribution is Gaussian:

$$N_A(r) = N_m \exp(-r^2/a_1^2), \tag{36}$$

where $a_1 = a$ in the case of total ionization; $a_1 = a\sqrt{2}$ in the case of weak ionization.

We will first consider the simpler case where the acceptors remain totally ionized at the detector room temperature (77 K). The electric field E_0 entrains the holes from the external sections of the cloud. The resulting negative space charge of the acceptors generates a field that retains the remaining holes. Assume the field E_0 is switched off (conserving the number of holes in the cloud). Then these holes occupy a spherical area of radius R. Expanding the exponent in (36) in terms of r-R we find that the acceptor concentration for $r > R$ decays at a characteristic distance

$$L_a = a_1^2/2R. \tag{37}$$

We assume that L_a is significantly greater than the local Debye length L(R) (corresponding to the charged acceptor concentration N_a(R)) although is much less than R:

$$L(R) \ll L_a \ll R. \tag{38}$$

This inequality is confirmed by the estimates performed below. The transition layer between the quasineutrality region ($p \sim N_A$) and the space charge region ($p \ll N_A$) has a width of $\sim$ L(R), i.e., it is insignificantly narrow compared to the layer of thickness L_a in which the majority of the space charge is concentrated. In turn this layer is narrow compared to R. Consequently the total space charge q may be taken as concentrated in a sphere of radius R and equal to

$$q = -eN_A(R)4\pi R^2 L_a, \tag{39}$$

where e is the modulus of electron charge. The radial electric field generated by this charge is equal to

$$E(r) = q/\varepsilon r^2 = -E^*(R/r)^2, \tag{40}$$

where ε is the dielectric constant; E^* is the field (modulo value) near the boundary $r = R$. When $r < R$ the behavior of the potential $U(r)$ is determined by Boltzmann's distribution for the holes, i.e., $U(r) = \left(\frac{kT}{e}\right)\left(\frac{r^2}{a^2}\right) + \text{const.}$

Now assume that pulling field E_o is again switched on. The holes are shifted in the direction of E_o which polarizes the charges and produces a supplementary field $\tilde{E}$. The boundary of the hole region remains virtually equipotential; the potential difference is equal to kT/e within an order of magnitude and is negligible compared to E_o. Consequently the polarizing field $\tilde{E} \sim E_o$, i.e., the surface density of acceptor charge at the boundary of the hole region will change significantly. For this process the boundary of the hole region will shift over a distance L_a, i.e., insignificantly. Therefore the hole region remains spherical; polarization field $\tilde{E}$ is the as for a metallic sphere in an external field [33]; its radial component magnitude is equal to $2E_0(R/r)^3\cos\vartheta$, where the angle ϑ is recorded from the direction of the field E_o. The radial component of the pulling field itself is equal to $E_0\cos\vartheta$. The total radial field at the boundary $3E_0\cos\vartheta - E^*$ will be negative everywhere so the holes are contained at $r < R$. In this case the potential relief $U(r,\vartheta)$ determined by the expressions given above for the electric fields, has the saddle point at $\vartheta = 0, r = R_S$. Specifically, if E^* is close to $3E_o$, then $R_S - R = R\sqrt{(E^*/3E_o)-1}$, while the potential difference δU between the saddle point and the boundary of the hole region is

$$\delta U = 2E_0R(E^*/(3E_0 - 1))^{3/2}. \tag{41}$$

In fact the holes will continue to drift from the cloud (and E^* will continue to grow) until the height of the barrier $e\delta U$ reaches certain values of kT. According to the inequality $kT/e \ll E_oR$ this will occur when E^* is very close to $3E_o$. The equation $E = 3E_o$ subject to (37), (39) and (40) determines the radius of the quasineutral hole region R:

$$\frac{a_1}{R}\exp\left(-\frac{R^2}{a_1^2}\right) = 3\varepsilon E_0/2\pi e a_1 N_m. \tag{42}$$

Due to the weak (logarithmic) dependence of R on E_o and the cloud parameters we may substitute a value of $N_m \sim 10^{16}$ cm^{-3} and typical values of $E_o = 10^3$ V/cm, $a_1 = 6$ μm; we obtain $R/a_1 \approx 2$. The inequality $L_a \ll R$ (i.e., $2(R/a_1)^2 \gg\gg 1$) is satisfied qualitatively. The inequality $L(R) \ll L_a$ subject to equations (36), (42) and the familiar expression $L(R) = \sqrt{\varepsilon kT/4\pi e^2 N_A(R)}$ may be rewritten as $E_o \gg kT/3eL_a \approx$ V/cm which is satisfied.

The radial electric field around the hole region attracts holes drifting in the detector, so the impurity cloud is a trap for the holes. The lines of force of the electric field terminate at the boundary ($r = R$); these lines pass through the cross-section σ far from the trap. It follows from the law of conservation of electric field flux that $\sigma E_o = 4\pi R^2 E^*$. Substituting $E^* = 3E_o$ into this expression we obtain the desired formula for the effective cross-section of the trap:

$$\sigma = 12\pi R^2. \tag{43}$$

The derived results may be directly applied to the case where the acceptors are weakly ionized (before application of the field). The only difference is that now the impurity clouds are divided into a neutral acceptor zone for $r < R$ and a completely ionized acceptor zone for $r < R$ (where the holes that have abandoned the acceptor levels due to thermal ionization are entrained by the pulling field). The complete electric field confines the holes (whose concentration is small compared to N_A) in the region $r < R$; at the boundary of this region the Fermi level of the holes coincides with the acceptor level of the impurity, i.e., the boundary is near-equipotential. Hence when inequality (38) is satisfied the former expressions are valid for electric fields when $r > R$, i.e., the formula for R and σ. The auxiliary condition for such a trap is that the thermal release time of the holes from the acceptor level will be small compared to the detector operating time, i.e., the acceptor level cannot be deeper than 0.2 eV.

If the acceptor level is deeper, the cloud remains neutral. In this case the drifting electrons pass through the cloud and are partially trapped by the neutral acceptors, i.e., the cloud is a macroscopic trap to the electrons. In order to determine the specific mechanism of charge losses in the detectors, a number of detectors fabricated from material that was first investigated by small-angle light scattering and selective etching techniques was also investigated by the amplitude charge transfer technique noted above.

The combined nonequilibrium charge was generated by γ-quanta Cs-137 isotope radiation with an energy of W_0 = 661 keV. Each quantum generated $\sim 2.3 \cdot 10^5$ electron-hole pairs in an initial volume of approximately 10^{-6} cm^3. The electron and hole packets were then divided by the electric field and drifted to the n^+ and p^--contacts, respectively. Due to the low absorption coefficient the generation points of the pairs are distributed with equal probability throughout the volume of the structure. During the drift process the carriers may be trapped, which is reflected in the charge amplitude Q recorded by capacitor plates equivalent to the structure. Each carrier trapped at a distance of x' from the electrode to which it drifts produces a reduction in Q of ex'/S. If $\Psi(x')dx'$ is the portion of initial holes absorbed at a distance from x' to $x' + dx'$ from the cathode, the contribution of absorbed holes to the relative reduction in accumulated charge λ is

$$\lambda = (Q_0 - Q)/Q_0 = \int_0^x \frac{x'}{S}\,\Psi(x')dx', \tag{44}$$

where Q_0 is the maximum accumulated charge (without carrier losses); x is the coordinate of the γ-quantum track. Trapped electrons make an analogous contribution to λ.

By using the amplitude analysis technique we may record the amplitude with a relative accuracy better than 10^{-4}, i.e., measure very minor trapping losses and obtain a detailed loss distribution function $dp/d\lambda$ (the amplitude spectrum). The integral loss characteristic is the quantity $\bar{\bar{\lambda}}$ averaged over a large number of pulses. Normally the experimental amplitude spectrum is represented as the observed quantum energy distribution $W = W_0 Q/Q_0 = W_0(1 - \lambda)$ and is characterized by the resolution δW equal to the spectral width at half maximum amplitude. Often the spectrum is triangular (Figure 41) corresponding to the distribution

$$\frac{dp}{d\lambda} = \frac{2}{3\bar{\lambda}}\left(1 - \frac{\lambda}{3\bar{\lambda}}\right),\ 0 < \lambda < 3\bar{\lambda} \tag{45}$$

In this case the parameters $\bar{\lambda}$ and σW are related by the simple relation

$$\delta W = \frac{3}{2} W_0 \bar{\lambda}. \tag{46}$$

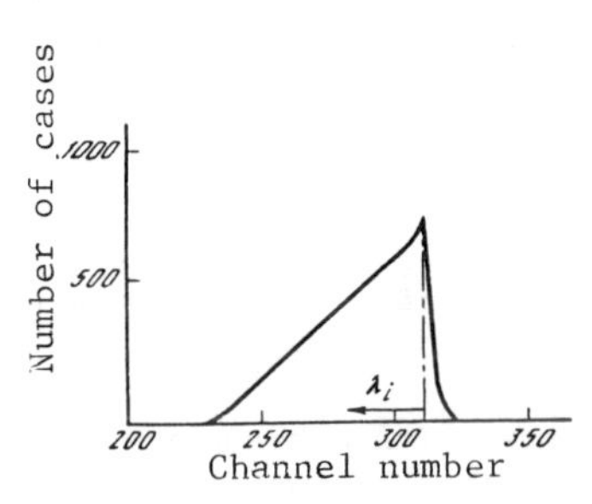

Figure 41. Spectral waveform of the amplitudes for sample No. 12.

The washed out right edge is caused by equipment noise and ionization fluctuations and on the left by hole trapping.

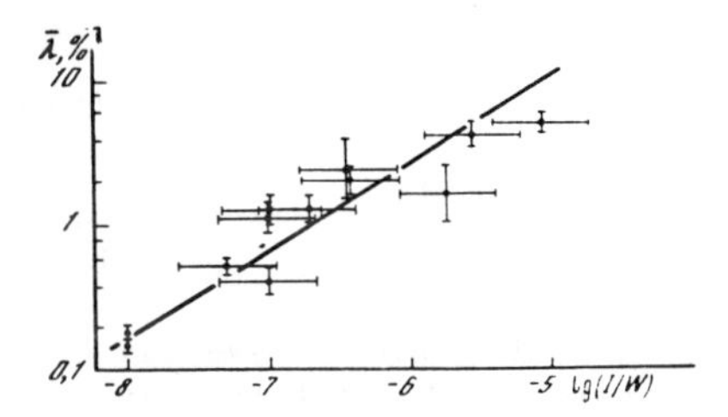

Figure 42. Mean losses in nonequilibrium charge transport plotted as a function of scattering intensity.

For the test group of samples (see Table 4) the quantity δW varies over a wide range: From 1.2 keV (which is close to the limit determined by ionization fluctuation) to 43 keV. When $\delta W > 43$ keV the measurements are complicated due to the merging of the loss spectrum with the Compton electron spectrum. A comparison of data on mean losses in the detector λ and small-angle scattering intensities for the initial samples I demonstrated that $\bar{\lambda} \sim I^{0.6}$ (Figure 42). The width of the amplitude spectrum δW has an analogous dependence on I.

In order to determine which carriers are subjected to preferential trapping, measurements were conducted on a number of samples with high losses in electron packet transport conditions. For this test the structures were irradiated on the p^+-contact side by Bi^{207} isotope β-particles with an energy of 972 keV and a path of ~ 0.8 mm. The derived spectra are significantly more narrow than the spectra of the γ-quanta and have $\bar{\lambda} \leqslant 0.4\%$. Consequently holes are preferentially trapped in pure Ge. We will discuss the possible trapping mechanisms.

1. *Trapping of holes by impurity centers.* The portion of trapped holes $\Psi(x')dx'$ is equal to $dx'/v\tau$, where v is the drift velocity, τ is the hole lifetime. It follows from (44) that $\lambda(x) = x^2/2v\tau S$, and

$$\bar{\lambda} = S/(\sigma v \rho). \tag{47}$$

The lifetime τ was measured by the photoconductivity kinetics; surface treatments provided a low surface recombination rate (less than 100 cm/sec). The derived values of $\tau \approx 20$ μsec correspond to losses of $\bar{\lambda} \approx 0.03\%$ which are significantly lower than the observed losses (see Table 4). Moreover, such small losses cannot have an influence on the actual amplitude spectrum, since the maximum width δW caused by ionization fluctuations amounts to 0.8 keV corresponding to $\bar{\lambda} = 0.12\%$. Another argument against explaining the losses as the trapping of impurity centers is related to the waveform of the amplitude spectrum $dp/d\lambda$. We find from the relations $\lambda(x) = 3\bar{\lambda}\, x^2/S^2$, $dp = dx/S$ [18]

$$dp/d\lambda = \frac{1}{2}(3\bar{\lambda}\lambda)^{-1/2}, \; 0 < \lambda < 3\lambda. \tag{48}$$

This spectrum having an integrated singularity at $\lambda = 0$ is significantly different from that given in [56].

2. *Hole trapping by charged acceptor clouds.* The average portion of trapped holes $\Psi(x')dx'$ (with a fixed coordinate of the track x) is equal to $\sigma C dx'$, where σ is the trap cross-section (determined by expression (43)), C is the trap (cloud) concentration. It follows from (44) that $\lambda(x) = \sigma C x^2/2S$. Averaging over x, we find

$$\lambda = \frac{1}{6}\sigma N S. \tag{49}$$

Based on the known values of $\bar{\lambda}$, S and the value $\sigma = 5 \cdot 10^{-5}$ cm^2 calculated from (43) we may find from (49) the trap concentration N; for samples with poor resolution ($\bar{\lambda} \sim 4\%$) we obtain $C = 10^4$ cm^{-3}.

Therefore in order to explain the observed losses we require a rather low trap concentration, i.e., a small total number of acceptors in all regions per unit of crystal volume ($\geqslant 10^{11}$ cm^{-3}). However it is difficult to fit this model to other experimental factors. First the derived value of C is significantly less than the cloud concentration of $\sim 10^6 - 10^7$ cm^{-3} estimated by the etch pit density. Further, with small C a significant portion ξ of the hole packets pass through the detector without encountering any trap; accounting for the diffuse spreading of the packet we may estimate $\xi = 25\%$ when $C = 10^4$ cm^{-3}. Therefore the amplitude spectrum will contain a narrow peak at $\lambda = 0$ with the area under the curve comprising only a portion of the total spectral area. Actually although the peak is observed when $\lambda = 0$ (Figure 41) its area is very small and corresponds to a trap concentration $C \sim 10^6$ cm^{-3}. Finally it is difficult to explain the correlation between λ and I (Figure 42), since λ is weakly dependent on the acceptor concentration, while I is proportional to the squared local hole concentration.

3. *Hole trapping by neutral donor clouds.* If the clouds are clusters of donor centers and the donor level is sufficiently deep, the clouds remain neutral; the neutral donors trap the holes drifting through the cloud. The local lifetime of the holes is equal to $\tau(r) = (v_{cp}\sigma_0 N_Д(\tau))^{-1}$, where σ_0 is the trap cross-section of the hole by the neutral donor centers; $N_Д(r)$ is the donor concentration determined by expression (36); v_{cp} is the mean hole velocity close to the drift velocity v in a strong field. If the mean flux density is equal to J and the probability of hole capture by the cloud is small, in an element of volume dV, $JdV/v\tau(r)$ holes are trapped per unit of time. Substituting the expression for τ and $N(r)$ into this expression and carrying out volume integration of the cloud, we find its effective trap cross-section

$$\sigma = \pi^{3/2} d_f^3 \sigma_0 N_m \bar{v}/v. \tag{50}$$

The mean charge losses $\bar{\lambda}$ are determined by expressions (49), (50). Formula (50) is valid if the hole trap probability is small, i.e., if σ is significantly less than the geometric cross-section πa_1^2. If the trap probability of the cloud is close to 1, then $\sigma \sim \pi a_1^2$. Using the known quantities λ, a_1 and the inequality $\sigma \leqslant \pi a_1^2$, we may estimate the trap concentration at $N \geqslant 5 \cdot 10^5$ cm^{-3} which agrees with the observed etch pit density and the low peak in the amplitude spectrum (when $\lambda = 0$).

The mean losses $\bar{\lambda}$ are proportional to $N_m a^3 C$. If the donors remain weakly ionized at room temperature as well, the maximum electron concentration is proportional to $\sqrt{N_m}$, i.e., λ is proportional to $Cn_m^2 a^3$ at the same time that the scattering intensity I is proportional to $Cn_m^2 a^6$. Then the correlation between the parameters $\bar{\lambda}$ and I becomes quite clear (there is no exact proportionality between λ and I/a^3, which may be due to the significant error of the quantity I as well as the existence of several types of deep donors).

Therefore the most probable charge loss mechanism in the detector is hole capture by neutral donor centers forming clusters (clouds).

A number of studies appeared in the 1980s [58] demonstrating on the basis of a detailed analysis of the energy spectra of the detectors, the influence of the electric field on the spectra and the operating temperature of the detector (between 8-80 K) that the most probable charge loss mechanism is carrier capture by charged acceptor clouds. However the material used to fabricate the detectors employed in [58] was not examined using the small-angle light scattering technique and the parameters of the oxygen clouds are unknown.

It is possible that the composition of undetectable impurities may dominate the different possible charge loss mechanisms in different crystals depending on their growing conditions. However the strong influence of oxygen clouds detected in this study on the charge accumulation process in detectors fabricated from ultra-pure germanium is beyond doubt. It will most likely be possible to clarify this process as well as predict the degree of cloud influence on detector resolution by determining the ionization energies of the centers comprising the oxygen clouds (for example using the temperature dependencies of the scattering intensity or by investigating the influence of impurity illumination, see Chapter 1, § 1).

11. The Influence of Impurity Clouds on the Conductivity of Ultra-Pure Germanium

The large-scale clusters of electrically-active impurities (impurity clouds) described in the preceding chapter normally occupy a small portion of the volume and hence have only a slight influence on crystal conductivity. However since improving the degree of purity of the crystals will reduce the average concentration of free carriers throughout the crystal bulk n and will also increase ($\sim \sqrt{n}$) the Debye screening radius l, in the purest semiconductors the "effective" cloud size R may be significantly greater than the true cloud size a ($R = a \sqrt{n_{\text{обл}}/n}$, where $n_{\text{обл}}$ is the free carrier concentration in the cloud). In this case the clouds may have a significant influence on the crystal conduction in pure materials. The influence of the clouds will be significant if their conduction type differs from that of the sample itself since in this case the cloud is surrounded by a p-n-junction, i.e., a broad space charge region.

Probably the purest semiconductors today are ultra-pure germanium crystals used to manufacture nuclear radiation detectors. The carrier concentration in such crystals at T = 77 K is $\sim 10^{10} - 10^{11}$ cm^{-3} which may produce an increase in effective cloud size to $\geqslant 100$ μm. In this material the space charge regions may overlap around the clouds and create macro-regions with sharp depletion of free carrier concentrations, i.e., an anomalously high resistance ρ. This section reports results from experiments to investigate the correlation between changes in resistance at T = = 77 K (ρ_{77}) and light scattering intensity by impurity clouds through ultra-pure germanium.

The microwave method [59] was used to perform measurements of ρ_{77} with relatively high locality ($\sim$ 1 mm). The samples were scanned in $\sim$ 1 mm steps. Light scattering was investigated using the standard technique for such experiments at room temperature. The diameter of the probe spot was $\sim$ 2-3 mm and scanning was performed in $\sim$ 2.5 mm steps across the same areas as the measurements of ρ_{77}.

Ultra-pure vacuum-grown dislocation germanium samples were used (analogous to the description in § 8). Its particular feature was the existence of a sharp inhomogeneity in ρ_{77} including regions with $\rho > 10^5$ Ω · cm (Figure 43, a, b). Such high values of ρ_{77} cannot be attributed to the degree of crystal purification since the value of ρ_{77} does not exceed 10^4 Ω · cm in more pure Ge samples [60, 61]. This effect was observed in crystals where the volumetric carrier concentration n_{77} was less than $5 \cdot 10^{10}$ cm^{-3}. (The values of n_{77} in the wafer regions where ρ_{77} was at a minimum were used as the volumetric carrier concentration.) Among the samples in which ρ_{77} is rather uniformly distributed in the cross-section of the wafer were rather high-resistance samples with $\rho_{77} \sim 10^4$ Ω · cm and this demonstrates that the observed inhomogeneities in the ρ_{77} distribution are not related to the microwave technique of measuring conductivity.

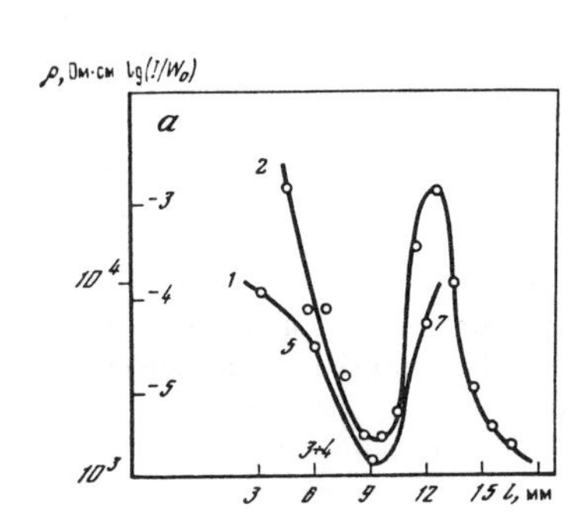

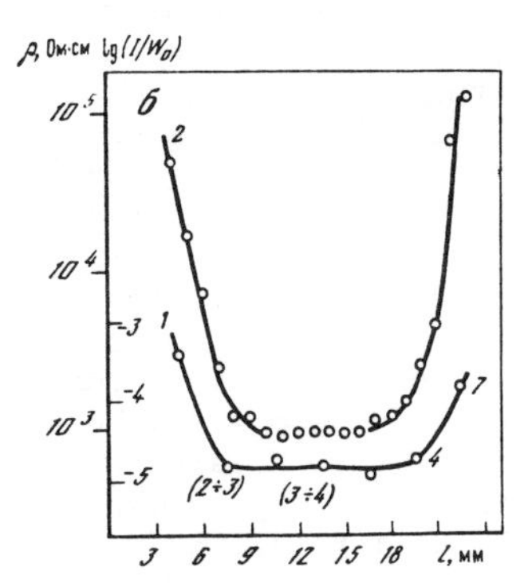

Figure 43. Sample scattering intensities (1) and conductivity (2) at T = 77 K plotted as a function of distance to the crystal edge I for two germanium crystals with an anomalous ρ_{77}.

The numbers by the various curve sections represent the dimensions of the impurity clusters in μm.

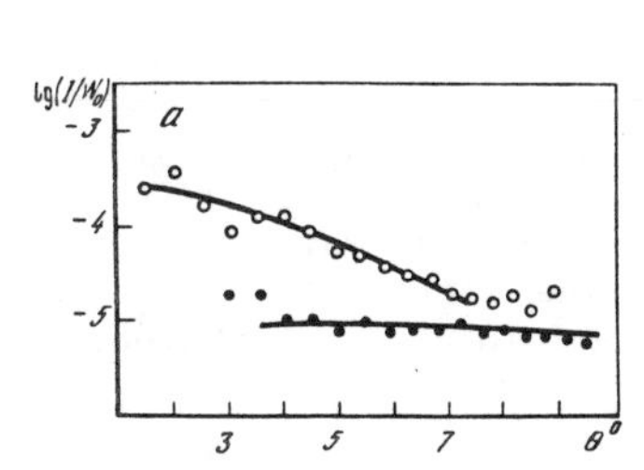

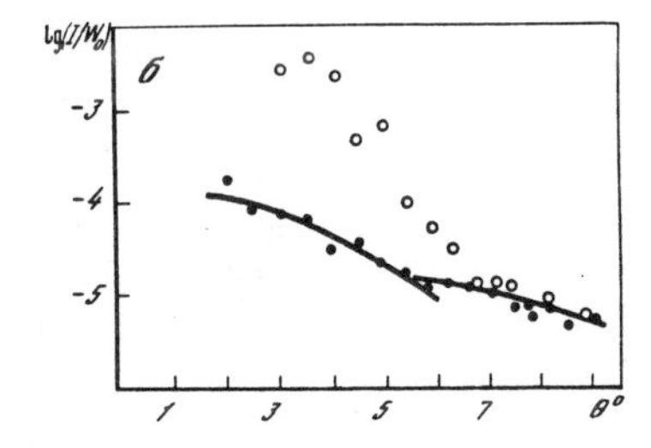

Figure 44. Characteristic scattering diagrams of ultra-pure germanium with anomalous values of ρ_{77}.

Figure 44, a, b, gives characteristic scattering diagrams of Ge samples with an anomalous ρ_{77} distribution. These diagrams clearly reveal Gaussian particles (impurity clouds) with $a \sim 5$, 7 and 9-10 μm where in a number of cases clouds on two scales are observed simultaneously (Figure 44, a). In a number of cases the entire diagram consists of plateau.

Occasional powerful light scattering by large inhomogeneities whose nature is unknown (Figure 44, b) is observed.

This study investigated seven samples with an anomalous ρ_{77} distribution. In the investigation of the samples a good correlation was observed between the scattering intensity by impurity clouds and the ρ_{77} values (Figure 43, a, b). Regions with $\rho_{77} \sim 10^3\ \Omega \cdot cm$ were characterized either by cloud scattering with $I_{pac} \leqslant 10^{-6}\ sr^{-1}$ or by a plateau with $I \sim 10^{-5}\ sr^{-1}$. Regions with $\rho_{77} \sim 10^4\ \Omega \cdot cm$ were characterized by cloud scattering with $I_{pac} \geqslant 10^{-5}\ sr^{-1}$ while in regions with $\rho_{77} \geqslant 10^5\ \Omega \cdot cm$ cloud scattering with $I_{pac} \geqslant 10^{-4}\ sr^{-1}$ was observed. No correlation was observed between scattering by large inhomogeneities and clouds of anomalous ρ_{77}. Occasionally their presence made it impossible to reliably identify cloud scattering. Naturally there are no such points in the $I(l)$ relations shown in Figure 43. Moreover light scattering is normally not successful for investigating points further than 5 mm from the edge of the sample. In addition we note that when clouds of two scales have been observed in the scattering diagrams the I_{pac} of the cloud type whose scattering was the most significant is plotted on the $I(l)$ relations.

We also investigated a crystal with a uniform ρ_{77} distribution. The scattering diagrams of this sample are in agreement with the scattering diagrams of regions with a minimum value of ρ_{77} and take the form of a plateau with $I \sim 10^{-5}\ sr^{-1}$.

Thus the derived results allow us to assume that the regions of anomalous ρ_{77} observed in ultra-pure germanium are caused by cloud overlap. Evidently even occupying a small portion of the volume they may in many cases have a significant influence on the electrical properties of the material.

The absence of a correlation between results from the investigation of ρ_{77} and light scattering for one of the samples may be related to the fact that when using the microwave method a rather thin surface layer is analyzed at the same time that when using the light scattering method the entire crystal bulk is analyzed (the thickness of the test wafers was ~ 5 mm). It is possible that the crystal bulk contains regions of anomalous ρ_{77} that are not registered when measuring ρ_{77} by the microwave technique.

BIBLIOGRAPHY

1. De Kook, A. J. R. Imperfections in Silicon Crystals. Philips Res. repts. 1973. Suppl. No. 1, pp. 1-105.

2. Ravi, K. V., Vacker, C. J. Stacking-Fault in Silicon, Induced by Oxidation. J. Appl. Phys., 1974, Vol. 45, No. 1, pp. 263-271.

3. Magden, I. N., Kalnes, N. A. Influence of Crystaliograthik and Electrophysical Imperfections in Monocrystalline Silicon on the Formation of Microplasmas in *p*-*n*-Junctions. Electron. Technol., 1975, Vol. 8, No. 3/4, pp. 33-43.

4. Stafeev, V. I., El'tsov, A. V. Recombination-Trap Barriers in Wideband Semiconductors. Fizika i tekhnika poluprovodnikov, 1976, Vol. 10, Issue 5, pp. 930-933.

5. Sheynkman, M. K., Shik, A. Ya. Long-Term Relaxations and Residual Conductivity in Semiconductors (a Survey). Fizika i tekhnika poluprovodnikov, 1976, Vol. 10, Issue 2, pp. 209-233.

6. Sheynkman, M. K., Markevich, I. V., Khvosmov, V. A. The Residual Conductivity Model in Semiconductors and Its Parameters in CdSe. Fizika i tekhnika poluprovodnikov, 1971, Vol. 5, Issue 10, pp. 1904-1911.

7. Gliechyk, K. D., Litovchenko, N. M., Merker, R. The Role of Impurities in the Formation of Quenched-In Recombination Centers in Thermally Treated Silicon. Phys. status solidi, 1976, Vol. 33, pp. K87-K90.

8. Usami, A., Fyjii, Y., Morioka, K. The Effect of Swirl Defects on the Minority Carrier Lifetime in Heat-Treated Silicon Crystals. J. Phys. D: Appl. Phys., 1977, Vol. 10, pp. 899-909.

9. Milevskiy, L. S. The Change in the Electrophysical Properties of Silicon Under the Influence of Point Defects in Dislocations. FTT, 1975, Vol. 17, No. 7, pp. 2200-2202.

10. Atsarkin, V. A., Mazel', E. Z. The Influence of Silicon Heat Treatments on the Lifetime of Nonequilibrium Charge Carriers. FTT, 1960, Vol. 2, No. 9, pp. 2089-2094.

11. Bel'skiy, I. G. Zakalochnye tsentry rekombinatsii v kremnii [Hardened Recombination Centers in Silicon]. Kremniy. Moscow: Izd-vo inostr. lit., 1960, pp. 340-345.

12. Reed, C. L., Mar, K. M. Gettering of Point Defects Within Oxygen Precipitates. J. Electrochem. Soc., 1981, Vol. 127, No. 9, pp. 2058-2067.

13. Mages, T., Loung, G. Gettering of Mobile Oxygen and Defects Stability Within Back-Surface Damage Regions. Appl. Phys. Lett., 1981, Vol. 38, No. 11, pp. 891-893.

14. Bagraev, N. T., Vlasenko, L. S. Electron-Nuclear Interactions in Silicon Under Optical Pumping. FTT, 1979, Vol. 21, No. 1, pp. 120-126.

15. Bagraev, N. T., Vlasenko, L. S. Opticheskaya polyarizatsii yadernykh momentov kak metod issledovaniya obrazovaniya magnitnykh primesnykh klasterov v kristallakh [Optical Polarizations of Nuclear Moments as a Method of Investigating Magnetic Impurity Cluster Formation in Crystals]. 6-ya Mezhdunar. konf. po rostu kristallov [The Sixth International Conference on Crystal Growing]. Moscow, 10-16 September, 1980. Rasshir. tez. dokl. Moscow, 1980, Vol. 4, pp. 285-286.

16. Bagraev, N. T., Vlasenko, L. S., Mashovets, E. S. Regions With an Elevated Impurity Concentration in Silicon Grown by the Czochralski Technique. Fizika i tekhnika poluprovodnikov, 1982, Vol. 16, Issue 11, pp. 2011-2017.

17. Konopleva, R. F. Elektrofizicheskie svoystva neuporyadocheniykh sistem [The Electrophysical Properties of Disordered Systems]. Prepr. Fiz.-tekhn. in-ta im. A. F. Ioffe, Vol. 670, Leningrad, 1980, 62 pages.

18. Eremin, V. K., Strokan, N. B., Tisnen, N. I. The Influence of Large-Scale Traps on the Properties of Semiconductor Nuclear Radiation Detectors. Fizika i tekhnika poluprovodnikov, 1975, Vol. 9, Issue 8, pp. 1575-1579.

19. Voronkov, V. V., Voronkova, G. I., Zubov, B. V. et al. Light Scattering by Microdefects in Si and Ge. FTT., 1977, Vol. 19, No. 6, pp. 1784-1792.

20. Voronkov, V. V., Voronkova, G. I., Zubov, B. V. et al. Scattering of Infrared Laser Emission: A Method of Investigating Local Inhomogeneities in Pure Semiconductors. FTT, 1981, Vol. 23, No. 1, pp. 117-126.

21. Rick, L. M., White, W. B., Spear, K. E. Average Speckle Size in Small-Angle Light Scattering. J. Cryst. Growth. 1975, Vol. 28, pp. 240-248.

22. Ryskin, V. S., Betekhnin, V. I., Slutsker, A. I. Detection of Submicroscopic Cracks in Stressed Rock Salt Crystals. FTT, 1973, Vol. 15, Issue 11, pp. 3420-3422.

23. Moriya, K., Ogawa, T. Direct Observation of Dislocation in Gap by Light Scattering. J. Cryst. Growth. 1978, Vol. 44, pp. 53-60.

24. Osvenskiy, V. B., Portnov, O. G., Grishin, S. P., Karpov, V. N. Eksperimental'noe issledovanie metodom rasseyaniya IK-sveta defektoobrazovaniya v monokristallakh arsenida galliya [Experimental Investigation of Defect Formation in Gallium Arsenide Monocrystals by Means of Infrared Light Scattering Technique]. Tez. dokl. 6-y konf. po protsessam rosta i sinteza poluprovodnikovykh kristallov i plenok. [Theoretical Papers of the Sixth Conference on Semiconductor Crystal and Film Growth and Synthesis Processes]. Novosibirsk, 21-25 June, 1982. Novosibirsk, 1982, Vol. 2, pp. 137-138.

25. Pokrovskiy, Ya. E., Svistunova, K. I. Light Scattering by Condensed Phase Nonequilibrium Charge Carrier Droplets in Germanium. Pis'ma v ZhETF, 1971, Vol. 13, Issue 5, pp. 297-302.

26. Sibel'din, N. N., Bagaev, V. S., Tsvetkov, V. A., Penin, N. A. An Investigation of Exciton Condensation in Germanium by Light Scattering Technique. FTT, 1973, Vol. 15, Issue 1, pp. 177-180.

27. Blistanov, A. A., Vishnyakov, V. I., Portnov, O. G. Rasseyanie sveta s λ = 1.15 mkm kristallami Si, vyrashchennymi metodom Chokhral'skogo [Light Scattering at λ = 1.15 μm by Si Crystals Grown by the Czochralski Method]. Tez. dokl. 3-y simpoz. po protsessam rosta i sinteza polyprovodnikovykh kristallov i plenok [Topic Papers of the Third Symposium on Semiconductor Crystal and Film Growing and Synthesis Processes]. Novosibirsk, 19-20 June, 1975. Novosibirsk, 1982, p. 273.

28. Van de Khyulst, G. Rasseyanıe sveta malymi chastitsami [Light Scattering by Minor Particles]. Moscow: Izd-vo inostr. lit., 1961, 623 pages.

29. Shiffrin, K. S. Rasseyanie sveta mutnymi sredami [Light Scattering sy Dense Media]. Moscow: Gostekhteorizdat, 1953, 237 pages.

30. Pertsev, A. P., Pisarevskiy, A. M. Odnozlektronnye kharakteristiki FEU i ikh primenenie [The Optoelectronic Characteristics of Photoelectron Multipliers and Their Application]. Moscow: Atomizdat, 1971, 218 pages.

31. Khird, G. Izmerenie lazernykh parametrov [Measuring Laser Parameters]. Moscow: Mir, 1970, 593 pages.

32. Landau, L. D., Lifshits, E. M. Elektrodinamika sploshnykh sred [The Electrodynamics of Continuous Media]. Moscow: Gostekhteorizdat, 1957, 620 pages.

33. Kostin, V. V., Kulevskiy, L. A., Murina, T. M. et al. Giant Pulses in CaF_2:Dy^{2+} Crystals. Prikl. spektroskopiya, 1967, Vol. 6, Issue 1, pp. 33-35.

34. Zubov, B. V., Murina, T. M., Olovyagin, B. R., Prokhorov, A. M. Investigation of Nonlinear Absorption in Germanium. Fizika i tekhnika poluprovodnikov, 1971, Vol. 5, Issue 4, pp. 636-640.

35. Vavilov, V. S. Deystvie izlucheniya na poluprovodniki [The Influence of Emission on Semiconductors]. Moscow: Fizmatgiz, 1963, 264 pages.

36. Zabolotskiy, S. E., Kalinushkin, V. P., Ploppa, M. G., Murina, T. M. An Assembly For Investigating Clusters of Electrically-Active Impurities in Semiconductors by the Small-Angle Light Scattering Technique. PTE, 1984, Issue 4, pp. 206-208.

37. Voronkova, G. I., Verbitskiy, N. I., Goncharov, L. A. et al. Sources of Impurity Cloud Formation in Germanium. Fizika i tekhnika poluprovodnikov, 1983, Vol. 17, Issue 12, pp. 2137-2142.

38. Voronkov, V. V., Voronkova, G. I., Murin, D. I. et al. A New Type of Impurity Center in Semiinsulating Gallium Arsenide. Fizika i tekhnika poluprovodnikov, 1984, Vol. 18, Issue 8, pp. 1363-1367.

39. Petroff, P. M., de Kock, A. J. R. Characterization of Swirl Defects in Floating-Zone Silicon Crystals. J. Cryst. Growth. 1975, Vol. 30, pp. 117-124.

40. Bernewits, H., Kolbesen, B. O., Mayer, P. Electron Microscope Study of Swirl Defects in Silicon. J. Appl. Phys. Lett., 1974, Vol. 25, No. 5, pp. 277-279.

41. Indenbom, V. L., Nikipenko, V. I. Napryazheniya i dislokatsiya v poluprovodnikakh [Stresses and Dislocations in Semiconductors]. Moscow: Izd-vo AN SSSR, 1961, 178 pages.

42. Khirt, Dzh., Loti, I. Teoriya dislokatsiy [Dislocation Theory]. Moscow: Atomizdat, 1972, 329 pages.

43. Reiger, H. Untersuchungen zum Ausscheidungsmechanismus von Kupfer in Silizium. Phys. Stat., sol., 1964, Vol. 7, pp. 685-689.

44. Boltaks, B. I. Diffuziya i tochechnye defekty v poluprovodnikakh [Diffusion and Point Defects in Semiconductors]. Leningrad: Nauka, 1972, 643 pages.

45. Milns, A. Primesi s glubokimi urovnyami v poluprovodnikakh [Impurities With Deep Levels in Semiconductors]. Moscow: Mir, 1977, 552 pages.

46. Roksnoer, P. J., Van den Boom, H. B. Microdefects in a Nonstriated Distribution in Floating-Zone Silicon Crystals. J. Cryst. Growth. 1981, Vol. 53, pp. 563-573.

47. Foll, H., Cosele, V., Kolbesen, B. O. The Formation of Swirl Defects in Silicon by Acclomeration of Self-Interstitials. J. Cryst. Growth. 1977, Vol. 40, pp. 90-108.

48. Seeger, A., Foll, H. Self-Interstitials, Vacancies and Their Clusters in Silicon. Intern. Conf. on Radiation Effects in Semiconductors. Dubrovnik, 1976. Bristol, London, 1977, pp. 12-29.

49. Veselovskaya, N. V., Sheykhet, E. G., Neymark, K. M., Fol'kevich, E. S. Defekty tipa klasterov v monokristallakh kremniya [Cluster Defects in Silicon Monocrystals]. Rost i legirovanie poluprovodnikovykh kristallov i plenok [Growth and Doping of Semiconductor Crystals and Films]. Novosibirsk: Nauka, 1975, pp. 284-288.

50. De Kock, A. J. R., Roksnoer, P., Boonen, P. G. T. The Introduction of Dislocations During the Growth of Floating-Zone Silicon Crystals as a Result of Defect Condensation. J. Cryst. Growth. 1975, Vol. 30, pp. 79-85.

51. Tempelhoff, K., Van Sung, N. Formation of Self-Disorder Agglomerates in Dislocation-Free Silicon During Crystal Growth. Phys. status solidi. 1982, Vol. 70, pp. 441-449.

52. Voronkov, V. V. The Mechanism of Formation of Swirl-Defects in Silicon. J. Cryst. Growth. 1982, Vol. 59, pp. 624-653.

53. Haas, G. The Diffusion of Oxygen in Silicon and Germanium. J. Phys. and Chem. Solids. 1960, Vol. 15, pp. 108-116.

54. Corbett, J. W., McDonald, R. S., Watkins, G. D. The Configuration and Diffusion of Isolated Oxygen in Silicon and Germanium. J. Phys. and Chem. Solids. 1964, Vol. 25, pp. 873-879.

55. Ryvkin, S. M. Fotoelektricheskie yavleniya v poluprovodnikakh [Photoelectric Phenomena in Semiconductors]. Moscow: Nauka, 1983, 343 pages.

56. Eremin, V. K., Dannengirsh, S. G., Strokan, N. B. et al. The Nature of Large-Scale Traps and Ultra-Pure Germanium. Fizika i tekhnika poluprovodnikov, 1977, Vol. 12, Issue 6, pp. 1191-1193.

57. Dannengirsh, S. G., Zatoloka, S. I., Kukushkina, T. I. Microdefects in Ultra-Pure Germanium. Izv. AN Latv SSR. Ser. fiz. i tekhn. nauk, 1983, No. 2, pp. 37-42.

58. Eremin, V. I., Strokan, N. B., Chikalova, O. P. An Investigation of Impurity Clusters in Pure Materials Based on Spectral Waveforms From the Innilation of Electron-Positron Pairs. Fizika i tekhnika poluprovodnikov, 1984, Vol. 18, Issue 2, pp. 244-249.

59. Tyul'kov, S. G. Issledovanie provodimosti osobo chistogo germaniya SVCh-metodom [Investigation of Conductivity of Ultra-Pure Germanium by the Microwave Method]. Tez. dokl. 2-y Vsesoyuz. shkoly PPD v yadernoy fizike [Topic Papers of the Second All-Union Conference on Semiconductors in Nuclear Physics]. Piga, 13-18 March, 1985. Moscos: Nauka, 1985, p. 46.

60. Haller, E. E., Joos, B., Falicov, L. M. Acceptor Complexes in Germanium: Systems With Tunneling Hydrogen. Phys. Rev. B. Solid State. 1980, Vol. 21, No. 10, pp. 4729-4739.

61. Haller, E. E., Hansen, W. L., Golding, F. S. Physics in Ultra-Pure Germanium. Adv. Phys., 1981, Vol. 3, No. 1, pp. 93-138.

Defect Formation by Intense Optical Excitation of Alkali-Halide Crystals with Mercury-Like Ions

V.P. Danilov

Abstract: Results are given from experimental investigations of intense UV laser irradiation on mercury-like ion-activated alkali-halide crystals. The effective photoionization of the impurity centers accompanied by the formation of high concentrations of defects in the crystals is investigated. The resulting defects are investigated by means of optical techniques and electron paramagnetic resonance. The photoionization cross-sections of the lower excited state of the impurity ions are measured and the quasilocal states of the mercury-like ions in alkali-halide crystals are discovered and investigated.

INTRODUCTION

Alkali-halide crystals activated by mercury-like ions (Tl^{+}, In^{+}, Pb^{2+}, etc., [1-3]) have been the focus of interest for more than a half-century (beginning with the studies by Hilsch and Pohl [4, 5]). One of the most important practical applications of such crystals has been and remains their use as ionizing radiation displays: Scintillators [6, 7]. They are widely employed as the working medium of lasers [8] and computer memories [9]. Pure and activated alkali-halide crystals have served for many years as model objects for the physics of crystal defects, luminescent center spectroscopy and other fields of solid state physics and physical optics. The absorption spectra that, as a rule, lie in the IR and the spectral-kinetic luminescence characteristics of alkali-halide crystals with mercury-like ions have been investigated in detail by regular spectroscopic techniques at low excitation densities [1-3, 10-17].

The present study is devoted to an investigation of the processes occurring in crystals under intense optical excitation ($I \geqslant 10^6$ W/cm^2). Such processes include the photoionization of impurity centers, photoconductivity, optical transitions from excited states and the formation of color centers.

The photoionization processes of impurity centers in wideband dielectrics (unlike the ionization processes in semiconductors and gases) represent virtually an ignored field of the physics of the interaction of radiation with matter. Investigating the photoionization processes of the activator yields information on the features of the band structure, the position of the quasilocal levels is the conduction band of the matrix and makes it possible to investigate the influence of the crystal field on the impurity ions.

A significant field of research on activated alkali-halide crystals is devoted to investigating defects that form in crystals under the influence of ionizing radiation (X-ray and gamma-quanta, and electron beams). However the defect formation processes resulting from optical excitation have been ignored.

All these data are important for many scientific and applied areas of quantum electronics and solid state physics dealing with the design of lasers employing crystalline defects, the laser stability of the optical materials and optical data recording.

The purpose of the present study is an experimental investigation of the processes in the effective ionization of mercury-like ions we discovered to exist in alkali-halide crystals under intense UV excitation. This problem became accessible due to progress in quantum electronics: The development of excimer lasers and access to the harmonics of crystal lasers.

Prior to this study research has appeared devoted to the influence of intense neodymium and ruby laser emission on activated alkali-halide crystals [18-20], i.e., in the transparency range of the crystals. The photoconductivity observed in this case was attributed by the authors of these studies to multiphoton excitation and ionization of the activator (In, Tl), with an ionization probability of 10^{-2} - 10^{-4} sec^{-1} [20].

We were the first to carry out experiments devoted to intense XeCl-laser (λ = 308 nm) emission action on alkali-halide crystals activated by mercury-like ions (Tl^{+}, In^{+}, Sn^{2+}, etc.). The experimental results revealed that intensive excitation in the longwave absorption band of the impurity ions (the A-band) results in the development of high concentrations of color centers (Tl^{+}, Tl^{o}-centers) [21-24]. A single excitation pulse (energy $E \approx 10^{-2}$ Joules and duration $\tau \sim 10^{-8}$ sec) generates in the crystal 10^{15} - 10^{16} cm^{-3} impurity color centers at an activation concentration of $C \sim 10^{18}$ cm^{-3}. The probability of ionization of the mercury-like ions by intense UV-excitation ($I \geqslant 10^{6}$ W/cm^{2}) is 10^{6} - 10^{7} sec^{-1} [22, 23] which is several orders of magnitude higher than the probability of ionization of the activator by intense ruby and neodymium laser emission.

One of the problems addressed in the present study was establishing the mechanism of the effective ionization of mercury-like ions in alkali-halide crystals under intense excitation in the A-band (4-5 eV). It has been assumed over the last two decades that optical excitation in the A-band will not ionize the activator nor lead to photoconductivity in crystals [25-27]. There has been evidence indicating that irradiating NaI-Tl crystals by a mercury lamp in the A-band will result in retention of luminescence [28], although this phenomenon has not been investigated in detail.

Our systematic investigations using a variety of techniques (pulsed absorption spectroscopy, luminescence methods and EPR) have revealed that the mechanism responsible for defect ionization of impurity mercury-like ions in alkali-halide crystals under intense excitation in the A-band is stepwise ionization of the activator from the lower excited state [22-24, 29]. It was also established that unlike X-ray bombardment, optical excitation to the activator absorption bands results in preferential formation of impurity centers and their concentration is at least an order of magnitude higher than that of the intrinsic defects (F-centers) [23, 24].

This study concentrates on an investigation of the optical transitions between the excited local states of the activator and the band states of the crystals. The fundamental focus is on measuring the cross-sections of the photoionization of the impurity ions from the excited states. The photoionization cross-sections are always important fundamental parameters that are needed to understand many processes in gases and condensed media. The experimental data needed to test theoretical models describing photoionization and for evaluating the promise of using activated alkali-halide crystals as operating media in lasers and optical data recording devices are particularly valuable [9, 30]. Theoretical estimates of the photoionization cross-sections of the excited states of the activator ions in alkali-halide crystals have been significantly complicated by possible distortions to the wave functions of the band states in the defect region [17, 31]. Therefore one of the first tasks formulated and performed for this class of crystals for the first time was measuring the cross-sections of the phototransitions between the excited states of mercury-like ions and the continuum states [29, 32].

Other subjects of indisputable interest also included investigating the spectral dependence of the photoionization cross-sections and determining the nature of the band states, i.e., the final states of the optical transition leading to ionization. This problem is related to the very interesting problem of investigating the quasilocal states of the activator: The unique analogs of the self-ionized states of the free atoms and ions [33, 34]. The situation when the excited states of the impurity fall within the continuum of the crystal-matrix is common: The quasilocal states are related to the L-bands of the F-centers [14, 17] and certain transitions of the F'-centers [35] and the Tl^{o}-centers [36]. The quasilocal impurity states have been discovered in a number of semiconductor compounds [37-39]. The quasilocal states of mercury-like ions were investigated in alkali-halide crystals in the early 1960s [12, 17, 31] although these studies were semi-phenomenological and theoretical and were not developed further. The difficulty of experimental investigations of the quasilocal states of mercury-like ions in alkali-halide crystals is that absorption to these levels from the ground level lies in the vacuum UV and normally overlaps with the exciton or fundamental absorption of the matrix [12]. Hence it is difficult to detect the quasilocal levels using regular "linear" spectroscopy.

Experimental data on the quasilocal impurity states are highly valuable for a wide variety of fundamental and applied studies. The energy position of the quasilocal states makes it possible to investigate the influence of the crystal field of the matrix on the activator levels and to investigate the genetic link between the levels of the activating atoms and ions in the free state and in the crystal. The quasilocal levels may have a significant influence on the photoionization processes of the impurity centers and in this aspect are of interest for optical data recording. The extent of the distribution of quasilocal states in activated dielectrics and semiconductors is evidently broader than would be anticipated from the many studies devoted to them. We believe that theoretical calculations and interpretations of such phenomena as the photoionization of the impurity centers and the impurity-trapping of carriers in actual systems are quite approximate if the possible quasilocal states are ignored.

The present study was the first to investigate the spectral dependence of the photoionization cross-sections of mercury-like ions in alkali-halide crystals from the excited states. These studies were carried out using two independent and complementary linear spectroscopy techniques: Absorption spectroscopy based on the spectra of our detected absorption from the excited state [40-43] and luminescence nonlinear spectroscopy based on the change in the relative quantum efficiency of the intracenter luminescence when the impurity centers are ionized by an excimer laser (λ = 308 nm) and the fourth harmonic from an Nd:YAG-laser (λ = 266 nm), i.e., quanta of different energies [44]. The photoionozation resonances detected in this case, as demonstrated by the analysis in the study, are most likely related to the quasilocal states of the mercury-like ions [40-43].

The fundamental results of the study may be used in developing solid state U V lasers which have applications in many scientific and applied laser spectroscopy fields [45, 46]. The preferential generation of impurity centers is of interest from the viewpoint of accumulating high concentrations of color centers in the crystals where the structure of the centers, as we have established by EPR studies [24], is analogous to the structure of Ag_F^0-centers [47]. These and similar centers are used as the active centers in IR-tunable lasers [8, 48]. The phenomenon of effective defect generation under optical excitation and the induced absorption in the crystals may be used as an optical data recording method.

CHAPTER 1

COLOR CENTER FORMATION BY INTENSE OPTICAL EXCITATION

1. Mercury-Like Ions in Alkali-Halide Crystals

The heavy metal ions Tl^+, In^+, Pb^{2+}, Sn^{2+} represent the traditional class of activators for alkali-halide crystals. The configuration of the optical electrons of these ions in the ground state is ns^2; the neutral mercury atom has an analogous configuration and hence the In^+, Tl^+ ions and other ions are often called mercury-like ions [1-3] (the foreign scientific literature also occasionally refers to them as Tl^+-like ions). The impurity mercury-like ions, as a rule, displace the metal cations at the sites of the alkali-halide crystal lattice, thereby forming luminescence centers whose optical properties have been extensively studied [1-3, 10-17]. When introducing impurity mercury-like ions in the alkali-halide crystals, broad absorption bands appear in the transparent region of the primary material (so-called A-, B-, C-, D-bands). In accordance with Seitz' model [49] a relation is established between the absorption bands and the ^1S_o - - $^3P_1$ (A-band), ^1S_o - $^3P_2$ (B-band) and ^1S_o - $^1P_1$ (C-band) electron transitions in the free ion. The D-band is normally related to the excitation of the near-activator exciton [12, 17]. At room temperature the $^3P_1$ - ^1S_o transitions predominate in the emission spectra [3, 12]. At low temperatures the $^3P_2$ - ^1S_o and $^1P_1$ - ^1S_o transitions may be found in the emission spectra, i.e., the adjacent B- and C-absorption bands [10, 50, 51].

The crystalline lattice field shifts the electric levels [1, 12, 49, 52], eliminates the degeneration of the electron states and changes the probabilities of the optical transitions [15, 52]. Due to the interaction of the optical electrons with the lattice vibrations the spectral absorption and emission lines are transformed into broad bands, i.e., the optical transitions of the mercury-like luminescence centers in alkali-halide crystals are electron-vibrational transitions [1, 52] for which an adiabatic approximation is used [53]. The interaction between the impurity ion and the ligands is easily described by means of potential curves that are successfully used to evaluate the most important characteristics of the spectra (the half-width of the bands, the Stokes shift, etc.) [1, 11].

Experimental investigations of the $^3P_1$-state (the $^3T_{1u}$-state in terms of the irreducible representations of the O_h point group) have revealed that it has a complex energy and spatial structure resulting from the interaction of the optical impurity electrons with the non-totally symmetric vibrations of the ligands (the Jahn-Teller effect) [14-17, 54-59]. The theoretically-predicted possibility for the coexistence of three types of equilibrium Jahn-Teller distortions to the ion configuration corresponding to tetragonal, orthorhombic and trigonal symmetries is implemented for mercury-like luminescence centers. Each group of equilibrium Jahn-Teller minima is a source of its own A-luminescence subband where there are generally three such subbands for the mercury-like centers (two proximate shortwave and one long-wave band). A Boltzmann distribution of the minima is established at high temperatures and emission from the lower (trigonal) minima (A_x-emission) predominates [58]. It is also established that below each minimum of the radiative state there are minima of identical symmetry of the metastable state $^3A_{1u}$ related to the ^3P_o level of the free ion of the activator [58, 59].

Alkali-halide crystals activated by mercury-like ions may contain various impurity and intrinsic defects formed during crystal growing and from the influence of ionizing radiation [60-65]. The most important intrinsic defects include the F-center and the V_K-center [14, 61-66]; these are the anion vacancy trapping the electrons and the Cl_2^- molecular ion localized at the two halide sites, respectively [14, 65, 66]. It has been established that the mercury-like ions, since they are luminescence centers, may function as electron and hole trap centers [67, 68]. The trapping of electrons and holes by mercury-like ions in the alkali-halide crystals results in the formation of electron A^{o} (In^{o}, Tl^{o}) and hole A^{2+} (In^{2+}, Tl^{2+}) color centers whose optical properties and structure were investigated in [36, 60, 62, 67, 68, 69-81]. Microwave-optical spectroscopy [81] has been used to perform a detailed investigation of defects in alkali-halide crystals (optical properties, structure and excited states).

Therefore the activator ions may influence the electron and hole distribution processes between the various trap centers. Such impurity color centers as A^{o}, A^{2+} are formed, as a rule, by generating in the crystal separated electrons and holes by irradiating the "band-band" transition range with X-ray and gamma-quanta emission. We know that irradiation in the C-absorption band may also produce free carriers due to the thermal ionization of the excited C-states of the impurity centers [25-27, 82, 83].

Laser excitation methods have made it possible to perform multiphoton ionization of the activator ions in alkali-halide crystals [18-20] and the ionization probability of the In^{+} and Tl^{+} ions in the alkali-halide crystals was varied over a range of $3 \cdot 10^{-2} - 3 \cdot 10^{-4}$ sec^{-1} for a ruby laser and $7 \cdot 10^{-4} - 2 \cdot 10^{-5}$ sec^{-1} for a neodymium laser [20]. For some time there were no reliable data on the ionization of the excited A-states of mercury-like ions, although the recombination luminescence observed in several experiments was interpreted as stepwise [28] or tunnel [84] ionization of the activator.

In recent years effective photoionization of mercury-like ions with the probability $W = 10^{6} - 10^{7}$ sec^{-1} has been detected for the case of intense excitation in the A-band [21-23]. Comprehensive investigations have experimentally confirmed the model of activator ionization from the excited A-state [22, 23] and the resulting color centers have been identified and their structure has been investigated [23, 24].

Nonlinear spectroscopy techniques have been employed to investigate the phototransitions between the excited and band states of the mercury-like ions in alkali-halide crystals and the cross-sections of such transitions have been measured [29, 32, 44] and the photoionization resonances have been detected; these have been attributed to the existence of the quasilocal states of the activator ions in alkali-halide crystals [40-43].

2. Experimental Set-Up and Spectral-Kinetic Investigation Technique

The primary methods used in the present study were optical methods to record absorption and luminescence as well as their spectra and kinetics and their dependence on the temperature, activator concentration, etc. In certain experiments the EPR of the crystal defects were investigated together with the photoconductivity and thermoluminescence. The optical studies were performed on the assembly whose block diagram is shown in Figure 1.

The excitation geometry was both longitudinal and transverse for investigating the absorption spectra by pulsed absorption spectroscopy and two-photon stepwise spectroscopy. An XeCl excimer laser (emitting at $\lambda = 308$ nm, pulse duration of $\tau \sim 1.5 \cdot 10^{-8}$ sec and pulse energy of E = = 10-50 mJ) was used as the excitation source together with the fourth harmonic from an Nd:YAG laser ($\lambda = 266$ nm, $\tau \sim 10^{-8}$ sec, $E = 5 \cdot 10^{-4}$ J).

All spectral-kinetic measurements were carried out on a double monochromator. The stationary optical absorption spectra of the crystals were recorded on the Specord and EPS-3 (Hitachi) two-beam spectrophotometers. Helium-neon, helium-cadmium and argon lasers as well as mercury lamps were used as the probe emission sources.

The FEU-39 and FEU-62 photoelectron multipliers were used to record the optical signals together with germanium photodiodes operating in both signal integration and temporal resolution modes. As a rule the oscillographic method of signal recording employing the S1-48B, S8-2 and C8-12 oscillographs was used in the spectral-kinetic investigations. In order to observe weak signals UZ-33 wideband amplifiers were included in the circuit between the emission detector and the oscillograph.

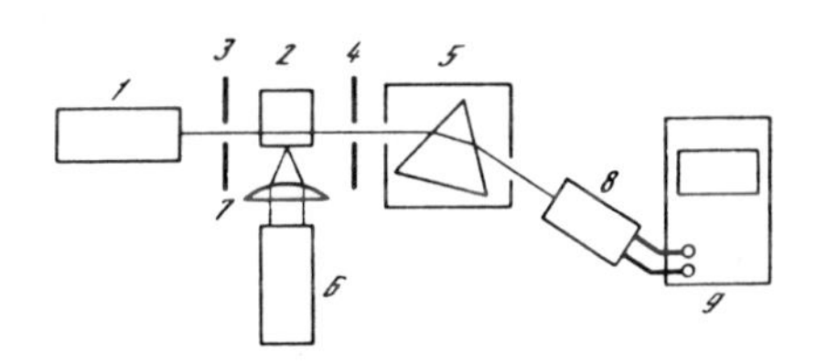

Figure 1. Block diagram of the experimental set-up.

1 - Probe light source; 2 - sample; 3, 4 - diaphragms; 5 - monochromator; 6 - UV laser; 7 - lens; 8 - photoelectron multiplier; 9 - oscillograph.

The high-sensitivity, broadband system developed for recording optical and electrical signals made it possible to detect and investigate many new physical effects in activated alkali-halide crystals (including anti-Stokes C-luminescence, and absorption from the excited state), as well as to expand the range of research and eliminate parasitic signals from the excitation sources. The EPR spectra were obtained on standard EPR-spectrometers (the JES-PE-3x) [81].

The method of investigating the photoresponse of the activated alkali-halide crystals was analogous to that developed and employed in [85, 86]. Capacitive coupling was used to couple the electrodes to the crystal through teflon plates (the ionization potential of teflon is ~ 10 eV), and the geometry of the test samples and the electrodes was selected to eliminate surface photoconductivity [85].

A metallic four-window cryostat with a cooled crystal holder (T = 90 - - 100 K) was fabricated for low-temperature research; a quartz cryostat with a liquid nitrogen crystal holder (T = 77 K) was also employed. The experimental assembly heated the test samples to 900 K. In all cases the temperature was measured by a graduated chromium-copal thermopile. A set of graduated light-filters was used to investigate the pump density dependences; these make it possible to vary the excitation intensity over 3-4 orders of magnitude.

The test crystals were grown from the melt by the Stockbarger and Kyropoulos methods with an activator concentration of $C = 10^{17} - 10^{19}$ cm^{-3}. The test samples were cleaved and polished, when necessary. In order to avoid aggregation of impurity ions which is quite prevalent for the double-charged ions (Sn^{2+}, Pb^{2+}) the samples were quenched immediately prior to the experiment by rapid cooling to 100-200°C below the melting point. Certain experiments employed ultra-pure alkali-halide crystals grown at the Institute of Crystallography of the Academy of Sciences of the USSR and the Institute of Physics of the Academy of Sciences of the Estonian SSR.

3. The Spectroscopy of Activated Alkali-Halide Crystals Under Intense Optical Excitation and Defect Identification

Prior to the present research the results of many experimentors [25-27, 82] had established that excitation in the A-band of alkali-halide crystals activated by mercury-like ions is not accompanied by recombination luminescence, photoconductivity nor the formation of impurity and intrinsic defects (with the exception of [28]), i.e., exciting the lower states of the activator does not result in their delocalization.

We discovered in our experiments that under intense excitation ($I \geqslant 1$ MW/cm^2) in the A-band of activated alkali-halide crystals there is a noticeable coloring of the crystals together with recombination luminescence that is damped over several seconds (at room temperature). When the probe beam is passed through the excitation region of the crystal, the light is absorbed during the pump pulse over a broad spectral range (0.3 - 0.7 μm) [21]. The induced absorption and recombination luminescence were observed under intense excitation in the A-band for a number of mercury-like ions in a variety of alkali-halide crystals: Tl^+ (KI, KCl, KBr), Sn^{2+} (KBr), In^+ (Kl, KBr, KCl, NaCl), Ga^+ (KCl, CsBr), Pb^{2+} (KCl, KBr, NaCl).

The thresholds for observing such effects varied from 10^6 to 10^7 W/cm^2 and the crystals for which excitation does not fall within the activator absorption band experienced no noticeable coloring. This fact observed in the initial simple experiments indicate that induced absorption and recombination luminescence are highly dependent on excitation photon interaction with the activator.

The range of investigations of crystals under intense UV excitation carried out in the present study includes absorption and luminescence measurements, EPR-spectroscopy, photoconductivity and thermoluminescence. The last two methods were not used to systematically investigate the crystals but rather these techniques were used as auxiliary methods for obtaining additional information.

The residual induced absorption spectra after irradiation by one or several laser pulses were recorded on two-beam spectrophotometers. For the crystals in which the induced absorption rapidly decayed (over a few seconds at T = 300 K including, for example, KI-Tl) the absorption spectrum was recorded by the probe light beam (see Figure 1) with the induced absorption kinetics recorded simultaneously on a storage oscilloscope. The optical density of the induced absorption was determined by the formula

$$D = \lg[I_0/(I_0 - I(t))], \qquad (1)$$

where I_o is the initial probe light intensity, $I(t)$ is the probe light intensity after the excitation pulse.

The induced absorption spectra after intense UV-irradiation in the A-band for KI-Tl and KCl-In crystals are given in Figure 2. A comparison of the derived induced absorption spectra to available data from identification of the absorption bands of defects in alkali-halide crystals [14, 71, 73, 81] has revealed that the spectral maxima in the long-wave region correspond to the absorption peaks of the A^o color centers (In^o, Tl^o). The induced absorption in the shortwave region of the spectrum is attributable to the hole color centers A^{2+} according to [21, 87]. The weak absorption of the F-centers is of interest; i.e., the intense irradiation in the A-band results in the preferential creation of impurity color centers.

The absorption coefficients of the defects estimated by the optical density D are 5-10 cm^{-1}. If we assume that absorption by the defects follows the allowed transitions (cross-section 10^{-16} cm^2), then in this case the concentration of generated impurity color centers is significant and amounts to $C \approx 10^{16}$ cm^{-3}. Therefore even a one time irradiation of the crystals with a laser pulse will generate significant concentrations of impurity color centers that absorb over a broad spectral range. The phenomonon of effective generation of defects absorbing over a broad spectral range (0.3-0.7 μm) discovered here may be used as an optical data transmission technique.

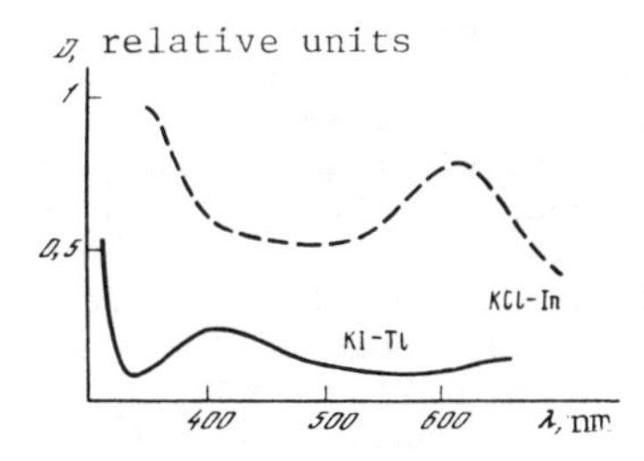

Figure 2. Induced absorption spectrum in KCl-In and KI-Tl crystals.

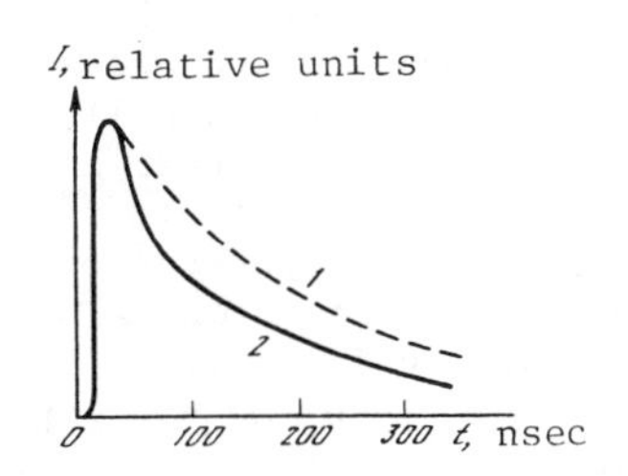

Figure 3. The luminescence kinetics of a KI-Tl crystal (λ = 430 nm) for two excitation levels: 10^5 W/cm^2 (1) and 10^7 W/cm^2 (2).

The intensity of the I curve is reduced by a factor of 60.

The spectral characteristics of the A-luminescence of mercury-like ions in alkali-halide crystals are virtually identical, according to the experimental results, when excited by low photon fluxes (an arc lamp) and under intense laser excitation. For certain crystals, particularly KI-Tl, Kcl-Pb [9], features have been discovered in the A-luminescence kinetics that are manifest in the knee of the spontaneous decay curve (Figure 3). The spectrum of the peak that appears for the case of intense excitation in KI-Tl does not coincide with the activator luminescence spectrum and is evidently attributable to the recombination emission from the uncontrollable impurities for recombination at the defects. Figure 3 indicates that increasing excitation intensity by two orders of magnitude will result in a 60 fold increase in luminescence intensity, i.e., the quantum efficiency of A-luminescence is reduced by a factor of 2. We investigated this phenomenon in detail (see Chapter 2) and it underlies the method of measuring photoionization cross-sections [29, 32].

In addition to activator luminescence, bands of various intensity and damping duration were also observed for the case of intense excitation in the A-band of alkali-halide crystals activated by mercury-like ions. Short-term ($\tau \sim 50$ nsec) luminescence was observed in a KCl-In crystal in the 570 nm range belonging to the In^{2+}-centers in accordance with studies [74, 88]. Analogous luminescence was discovered in KBr-In and KI-In crystals in the 550-600 nm range that also evidently belong to the In^{2+}-centers.

We detected weak anti-Stokes luminescence in the interval between the A- and C-absorption bands in KI-Tl and KI-In crystals. An interesting feature of anti-Stokes luminescence is that it has a clearly expressed quadratic dependence on the excitation intensity and investigating this phenomenon may yield valuable information on the mechanism of the observed ionization of the impurity centers.

Figure 4 presents the anti-Stokes luminescence spectrum in KI-Tl and the B- and C-luminescence spectra in this crystal based on data from [51]. It is clear that the detected luminescence may be interpreted as emission from the C-states of the activator, although this is not certain due to the possible manifestation of exciton luminescence in this spectral range [65].

In order to establish the nature of anti-Stokes luminescence we investigated a number of pure and activated alkali-halide crystals: KI, KBr, KI-Tl, KBr-In, KI-In. Pure alkali-halide crystals under intense excitation ($\lambda = 308$ nm) reveal luminescence whose spectrum and kinetics are in good agreement with the reliably identified emission bands of self-localized excitons [64, 65]. The luminescence from the singlet self-localized excitons in KI and KBr is anti-Stokes luminescence and has a square-law dependence on the excitation intensity. The latter fact clearly indicates two-photon excitation, since the bandgaps of the KI ($E_g = 6.3$ eV) and KBr ($E_g = 7.5$ eV) crystals are significantly greater than the pump quantum energy ($E = 4$ eV).

The existence of exciton luminescence in pure crystals suggests that it may also appear in activated crystals, although in the KBr-In crystal which has the optimum conditions for observing the luminescence of singlet self-localized excitons (the peak of the exciton emission exactly coincides with the absorption minimum between the A- and C-bands) and in other crystals (KBr-Sn, KCl-In) the anticipated luminescence was not observed. We may conclude based on results from these experiments that the anti-Stokes luminescence discovered in KI-Tl and KI-In crystals is attributable to emission from the B- and C-states of the impurity ions.

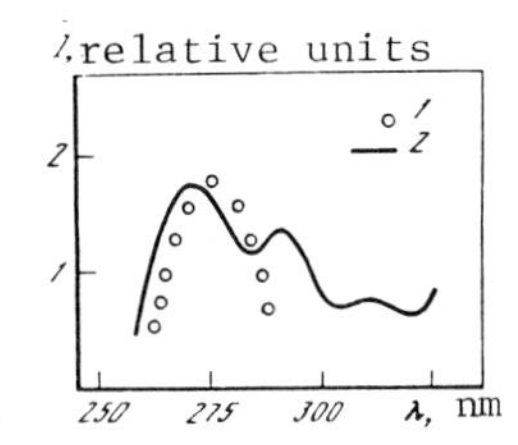

Figure 4. The anti-Stokes luminescence spectrum of a crystal under intense excimer laser excitation (λ = 308 nm).

1 - Experimental point;
2 - the B- and C- luminescence spectra according to [51].

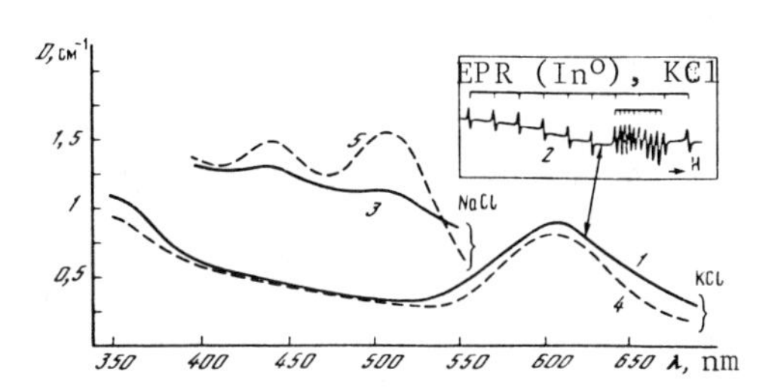

Figure 5. The optical absorption spectrum (1) and EPR spectrum (2) of a KCl-In crystal and the optical absorption spectrum of a NaCl-In crystal (3) under intense excimer laser excitation (λ = = 308 nm, τ = 15 nsec, E = $4 \cdot 10^{-2}$ J) at room temperature.

Curves 4 and 5 correspond to crystals irradiated by X-ray and F-light.

Since optical identification of defects is not entirely reliable in certain cases [81] in the present study we investigated the color centers that form under intense excitation in KCl-In and NaCl-In crystals using combined EPR and optical techniques. The EPR method was first used to investigate the defects generated in activated alkali-halide crystals under short-term optical excitation and made it possible to unambiguously establish the nature of these defects.

Figure 5 shows the optical absorption spectra of a KCl-In crystal irradiated by an XeCl laser (λ = 308 nm, I $\approx$ 10 mW/cm^2). An absorption band is observed (λ = 610 nm); this is consistent with that identified previously in [79] in X-ray irradiated crystals as (In^0)-centers. The EPR spectrum of the (In^0)-centers corresponding to optical absorption is shown in Figure 5 (curve 2). Ten lines are shown that may be attributed to the hyperfine interaction of an unevaporated electron and the nuclear magnetic moment. The linewidths are determined by superhyperfine interaction with the ligands.

Previous research [79] has revealed that the (In^0)-centers are evidently complexes of an In^+ ion and the adjacent anion vacancy that trap the electron, i.e., analogs of Ag_F^0 and Cu_F^0-centers [47, 89]. The unpaired electron is localized in the indium and partially (~ 30%) in the anion vacancy.

(In^o)-centers are also effectively generated in NaCl-In crystals irradiated by pulsed UV radiation (λ = 308 nm) in the A-band. Figure 5 shows the optical absorption spectrum (3) of an NaCl-In crystal after UV excitation at 300 K. Two absorption bands are observed that correlate with the EPR spectrum of the (In^o)-centers. Therefore the EPR method has revealed that intense UV irradiation in the A-band of the In^+ ions generates over time $\tau \sim 10^{-8}$ sec significant concentrations (10^{14} - 10^{16} cm^{-3}) of (In^o)-centers which evidently include an anion vacancy.

The majority of (In^o)-centers, as demonstrated in § 4, are formed over the laser pulse period of $\tau \sim 10^{-8}$ sec. Since the anion vacancy cannot reach the impurity ion of the indium in $\tau \sim 10^{-8}$ sec, we may assume that anion vacancies exist in the crystal prior to irradiation next to the In^+ ions. However such a mechanism is not likely since it would be difficult to assume the existence of a large quantity of vacancies (up to 10^{16} cm^{-3}) in immediate proximity to the activator before irradiation. It is more likely that due to the ionization of the impurity ions, near-activator excitation is produced that breaks down into an anion vacancy with an electron near the activator (the F-center at the nearest neighbor of the In^+ ion) and the Cl^o interstitial atom, i.e., a process occurs that is analogous to the creation of F and H-centers from exciton decay in the crystals [64].

The EPR spectrum of the activator color centers, similar to their optical absorption, is observed in KCl-In under excitation to the impurity absorption band. If excitation (λ = 308 nm) does not reach the impurity absorption band, as is the case for KCl-Tl and KCl-In (at nitrogen temperatures [16]), F-centers with a characteristic absorption band (2.3 eV) are preferentially formed in the crystals. Analogous effects have also been observed for KI-Tl [23] (see § 6).

4. The Kinetics of Defect Formation and Decay

The formation and decay kinetics of defects formed from pulsed UV excitation in alkali-halide crystals activated by mercury-like ions were investigated using the assembly described in § 2 (Figure 1) by means of pulsed absorption spectroscopy. At the point where the intense pump pulse is applied to the crystals, optical absorption appears over a broad spectral range and the absorption damping is not accompanied by spectral transformation (in the 20 nsec - 1 msec range).

Figure 6 shows the induced optical absorption spectrum in a KI-Tl crystal at initial time and 1 msec after application of the excitation pulse. It is clear that the spectral waveform and position of the maxima and minima in the spectrum do not change.

Figure 7 shows an oscillogram of the induced absorption pulse for an KI-Tl crystal in a spectral range of ~ 440 nm (HeCd laser). The oscillogram reveals fast and slow components. The slow component decays over ~ 0.2 sec while the fast component decays over ~ 0.7 msec (300 K).

The existence of fast and slow components in the induced absorption kinetics is also characteristic of other crystals. At room temperature the damping time of the induced absorption significantly exceeds the lifetime of the excited state of the activator ions. Later by virtue of the improved resolution and sensitivity of the assembly for certain crystals (KI-In, KBr-In, KBr-Sn) we detected absorption in the 300-400 nm range with a damping duration that coincides with the lifetimes of the excited states of the mercury-like ions. Chapter 3 is devoted to a detailed investigation of this phenomenon.

The damping kinetics of the induced absorption are largely dependent on temperature. The damping time of the fast component drops with an increase in temperature while the slow component remains unchanged. Since the principal electron trap centers in the test crystals are activator ions (see § 3), it is natural to assume that the damping kinetics of the induced absorption are determined primary by the Tl^o- and In^o-centers. Here the following situation is possible. With a low concentration recombination may occur without intermediate trapping and the recombination kinetics will be determined by the direct interaction of Tl^{2+} and Tl^o-centers. In this case the temperature dependence of the reduction in Tl^{2+}-center concentration may be represented in the following manner:

$$D = D_0 \exp[-(u/kT)], \tag{2}$$

where u is the temperature activation energy of the Tl^o-centers.

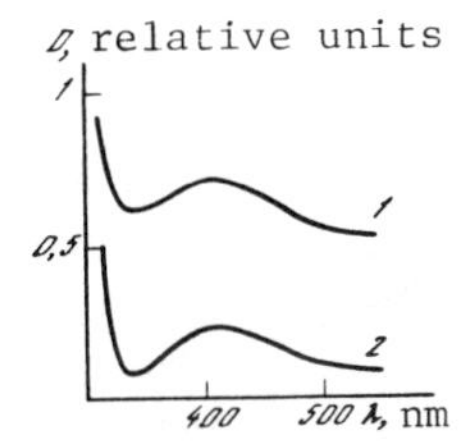

Figure 6. Optical density of induced absorption in an KI-Tl crystal.

1 - Maximum value at initial time; 2 - 1 msec after application of pump pulse.

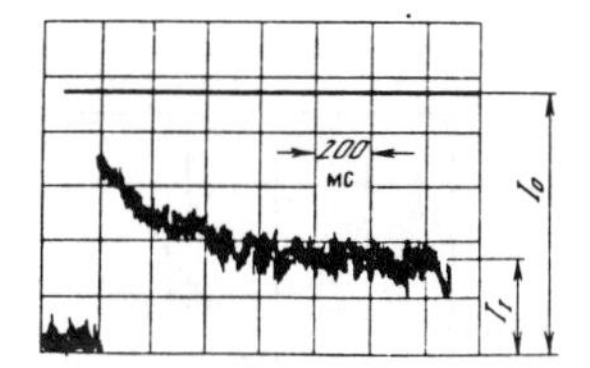

Figure 7. Oscillogram of induced absorption pulse in KI-Tl crystals.

As the activator concentration increases the probability of intermediate trapping of the electron before its recombination with the Tl^{2+}-center increases and the kinetics of the reduction in color center concentration in this case is complicated and is not described by expression (2). Experiments have revealed that at ion concentrations of $Tl^+ \sim 5 \cdot 10^{17}$ cm^{-3} virtually all the Tl^{2+}-centers decay exponentially with a characteristic reduction time in the optical density of $\tau \sim 0.7$ msec at T = 293 K. As the concentration of Tl^+ ions increases the slow damping component of the optical density appears and grows.

The temperature dependence of the fast component of the induced optical density becomes linear at the coordinates $\ln\tau - T^{-1}$ (see Figure 8). Here the activation energy is $u = 0.49 \pm 0.05$ eV, which is in good agreement with the thermal breakdown energy of the Tl^0-centers [90]. Consequently the fast region of variation in the optical density of the Tl^{2+}-centers is due to monomolecular recombination of the thermally-liberated electrons from the Tl^0-centers. The slow part of the variation in optical density is evidently attributable to bimolecular recombination of electrons liberated from the F-centers due to thermal and possibly tunnel processes.

The spectral absorption range of the activated electron color centers for certain crystals such as KCl-In reveals a small increase in the optical density after the steep induced absorption edge; this may be seen in Figure 9. This may be caused by the formation of additional (In^0)-centers by diffusion of the anion vacancies to the (In^0)-centers. We note that the creation of In^0-centers in KCl-In crystals under intense excitation in the A-band was fixed in the experiments not only by EPR and optical techniques but also based on the thermoluminescence peak at 353 K which is unambiguously related to the activator defect [81, 82].

The induced absorption edge for the case of pulsed excitation for all test crystals does not exceed the laser pulse duration. Figure 10 provides a sample induced absorption edge in a KI-Tl crystal for an HeCd laser probe signal (λ = 0.44 μm).

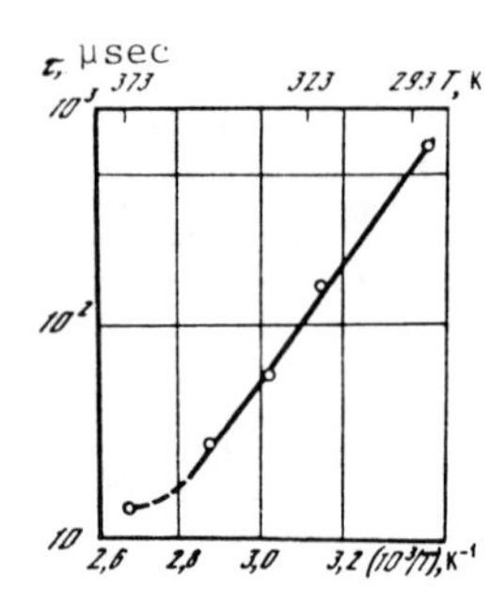

Figure 8. The temperature dependence of the damping time of the fast induced absorption section in a KI-Tl crystal.

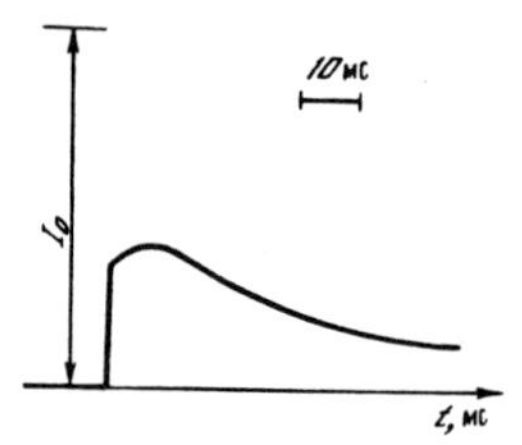

Figure 9. Damping of probe emission (λ = 632.8 nm) due to the absorption induced in an KCl-In crystal under pulsed excimer laser irradiation (λ = 308 nm).

As we will demonstrate below (see Chapter 2) the induced absorption edge contains important information that in many cases is decisive in discussing the ionization mechanism of mercury-like ions in alkali-halide crystals under intense excitation in the A-band.

We also investigated the kinetics of the ionization processes based on the photoconductivity of KBr-In, KI-Tl, KCl-In, etc., crystals. A typical photoconductivity pulse for the case of a KCl-In crystal is shown in Figure 11.

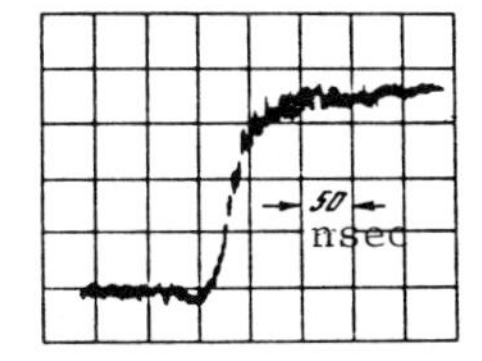

Figure 10. The induced absorption edge in a KI-Tl crystal (λ = 440 nm) for the case of pulsed excitation in the A-band.

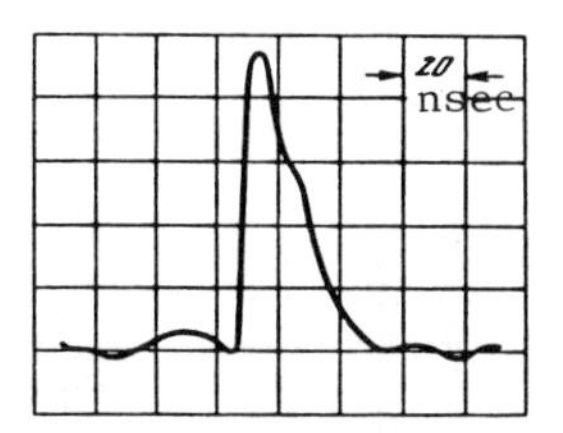

Figure 11. Photoconductivity kinetics in an KCl-In crystal under excimer laser excitation (λ = = 308 nm).

We know that in interpretating experiments devoted to multiphoton excitation of photoconductivity of both activated and pure alkali-halide crystals is normally complicated by uncontrolled impurities in the crystal that generate real intermediate levels in the conduction band [91]. Therefore we attempted to identify the photoconductivity that may be directly attributed to activator ionization. For this purpose we employed a method that has already been used in luminescence measurements, i.e., we investigated a set of activated and nonactivated crystals.

In order to eliminate possible manifestations of two-photon processes the experiments were carried out on KCl and KCl-In crystals since the sum of the two excitation quanta (2E = 8 eV) is less than the bandgap in KCl (E_g = 8.7 eV). In order to determine the influence of uncontrolled impurities a single nonactivated crystal sample of ultra-pure quality was a KCl crystal (1) while the other KCl crystal (2) was purified by multiple zone floating (at the Institute of Physics of the Academy of Sciences of the Estonian SSR). All crystals were of an identical size at 5 x 5 x 5 mm with an excitation intensity range of I = 10^6 - 10^7 W/cm^2.

The experiments revealed that the photoconductivity in nonactivated crystals was due to single-photon ionization of the uncontrolled impurities (the photocurrent amplitude is proportional to the photoexcitation intensity). The photoconductivity in KCl-In could be attributed most likely to the two-photon stepwise transitions, since the dependence of the photocurrent amplitude on I is almost a square-law relation. In spite of the existence of activator trap centers the photocurrent in the KCl-In crystal exceeds the photocurrent in KCl caused by uncontrolled impurities by a factor of 5-10.

On this basis we may conclude that the predominant part of the photoresponse in KCl-In is related to activator ionization. The photoresponse of the KBr-In, KI-Tl crystals and other crystals is analogous to the photoresponse of the KCl-In crystal and is also evidently caused by photon interaction with the activator. The kinetics of induced absorption and photoconductivity in activated alkali-halide crystals reveal that the impurity ions are ionized only during the laser excitation pulse.

CHAPTER 2

IONIZATION OF MERCURY-LIKE IONS IN ALKALI-HALIDE CRYSTALS UNDER INTENSE UV IRRADIATION

5. Ionization Probability

The research results presented in Chapter 1 allow us to assume that an effective mechanism exists for the generation of impurity color centers from the excitation of activated crystals in the A-band. In order to estimate the effectiveness of defect generation we require quantitative parameters of the ionization process, primarily the probability which may be determined in the following manner:

$$W_i = [N_{A^{2+}}/N_{(A^+)^*}]/\tau, \tag{3}$$

$N_A{}^2$ and $N_{(A+)*}$ are the concentrations of ionized and excited impurity ions; τ is the excitation duration.

The reason that our determination of the ionization probability is based on the number of excited impurity ions is that the effective ionization of the impurity ions is related specifically to their excitation. In certain cases we will use the degree of ionization, i.e., the ratio of the concentration of ionized activator centers to the concentration of excited centers. The number of excited ions may be determined from the expression

$$N_{(A^+)^*} = æ \int_0^\tau I(t)dt, \tag{4}$$

where æ is the absorption coefficient at the pump wavelength.

In order to determine the concentration of ionized activator centers we will use the familiar Smakula formula [92]. Dexter [93] has carried out a quantum mechanical calculation of impurity atom absorption in a solid and has derived a formula analogous to Smakula's formula:

$$N_f = 1{,}29 \cdot 10^7 [n/(n^2+2)^2] æ_{max} \Delta E, \tag{5}$$

where N is the concentration of absorbing centers; f is the oscillator force for the absorption band; n is the refractive index for the wavelength at the maximum of the absorption band; $\varkappa_{max}$ is the maximum absorption coefficient, cm^{-1}; ΔE is the half-width of the absorption band, eV.

Dexter demonstrated that the coefficient of 1.29 obtained for the Lorentz curve should be equal to 0.87 for a Gaussian curve, and the latter value is closer to our case [1]. The oscillator force for the A^{2+}-centers varies between 0.1-0.2 [78]. A simple calculation by formulae (3)-(5) reveals that the degree of ionization of the mercury-like ions (In^+, Tl^+) in alkali-halide crystals at $I = 10^6 - 10^7$ W/cm^2 is $10^{-2} - 10^{-1}$. This corresponds to an ionization probability of $W_i = 10^6 - 10^7$ sec^{-1}. These estimates reveal a high efficiency in the creation of impurity color centers by UV laser irradiation which significantly increases the effectiveness of the creation of such centers by X-ray irradiation [24].

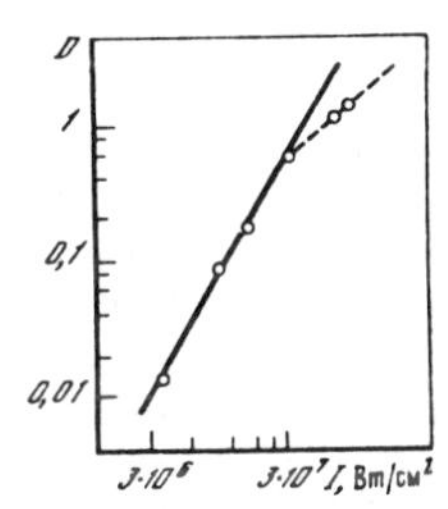

Figure 12. The optical density of induced absorption (λ = 440 nm) in a KI-Tl crystal plotted as a function of pump intensity.

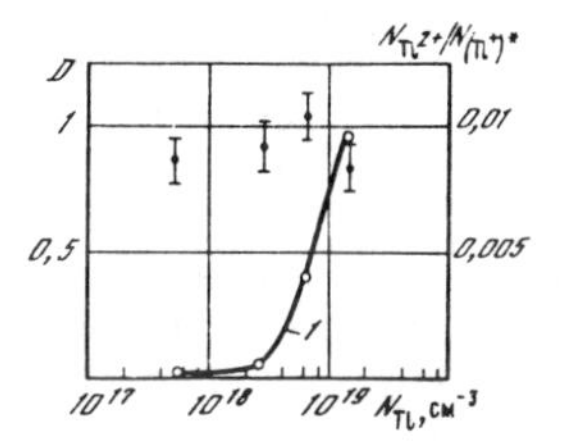

Figure 13. The concentration dependencies of the optical density of induced absorption (1) and the degree of ionization (the points indicate the confidence ranges) with a fixed pump intensity.

In order to determine the ionization mechanism which will be discussed in the next section we investigated the probabilities (degrees) of ionization as a function of activator concentration and excitation intensity. Figure 12 gives the experimental values of the optical density of induced absorption plotted as a function of pump intensity for a KI-Tl crystal (activator concentration $C \approx 10^{18}$ cm^{-3}).

The dependence of the optical density of color center absorption on excitation intensity manifests a nonlinear nature with a slope close to two, as we see from Figure 12 which is plotted in logarithmic coordinates. This is in agreement with results from investigations of the C-luminescence and photoconductivity in crystals and again indicates the two-photon nature of the ionization processes of impurity centers under laser irradiation. At an intensity of $I \geqslant 3 \cdot 10^7$ W/cm^2 the slope of the induced absorption relation changes, i.e., the degree of nonlinearity of the process is reduced. Possible causes of this phenomenon are discussed in § 6.

With a fixed pump intensity the optical density of induced absorption increases with growth of the Tl^+ ion concentration up to values of C = = $1.5 \cdot 10^{19}$ cm^{-3} (Figure 13).

An important relation from the viewpoint of establishing the ionization mechanism is the degree of dependence of ionization on the active particle concentration which, as noted by P. P. Feofilov [94] may yield an unambiguous criterion for differentiating the mechanisms in many cases. The degree of ionization of the Tl^+ ions in KI, as we see from Figure 13, is independent of the activator concentration within the confidence range. Subsequently (see § 7) the independence of the ionization processes of mercury-like ions in alkali-halide crystals of the activator concentration was confirmed by luminescence studies [32].

6. Ionization Mechanism

The purpose of this section is to discuss possible ionization mechanisms based on existing experimental facts and to establish an adequate ionization mechanism of mercury-like ions in alkali-halide crystals under intense excitation in the A-band.

According to the established classification, so-called lower ionization (trapping of holes by an impurity center) and upper ionization, i.e., the escape of the electron from the impurity center are possible [27, 83]. In the general case it is also possible that the action of intense UV radiation on the activated alkali-halide crystals will result, in addition to excitation of the activator, in two-photon generation of electron-hole pairs which, since they are trapped by impurity ions, may form A^{2+}- and A^{0}-centers. However for effective ionization ($W_i = 10^6 - 10^7$ sec^{-1}) of the activator the critical element is interaction between the excitation photons and specifically the impurity ions (see Chapter 1), rather than the matrix ions. Moreover, we know from experiments devoted to the influence of nanosecond electron pulses on KI-Tl and NaI-Tl crystals that lower ionization of mercury-like ions has a characteristic induced absorption time of the A^{2+}-centers in the order of hundreds of microseconds due to the migration of the V_k-centers to the trap centers [87].

In our examples the absorption edge of the A^{2+}-centers does not exceed 10^{-8} sec. Consequently the lower ionization mechanism will not play a significant role in the formation of A^{2+}-centers.

We will now consider possible upper ionization mechanisms. At present we know of several upper ionization mechanisms: Thermal ionization of an excited state [25-27], tunnel ionization [95] and photoionization [20, 27, 96, 97].

1. Thermal ionization. Previous investigations have revealed that thermal delocalization of electron excitations from the A-states is not observed when activated alkali-halide crystals are excited by low-intensity light sources (lamp excitation) [25-27]. The laser pulse action on the activated crystal may also result in local heating of the excited bulk. In

order to estimate local heating we will assume the following parameters: Energy per unit of volume: 1 J/cm^3, specific heat capacity: 0.29 J/(g·K) [98] and a density of 2 g/cm^3 [98]. Some simple calculations show that the temperature jump in the excited bulk does not exceed 400° even if all the absorbed energy is converted into heat. Hence thermal ionization of the A-state is not likely due to the large energy gap between the conduction band and the excited $^3P_1$ level (1 eV ~ 11600 K).

2. Tunnel ionization. Tunnel trapping of an electron from the excited C-state of the In^+ ion in KCl was observed in [95] which demonstrated that the probability of tunneling W_t between the luminescence centers in alkali-halide crystals is determined by the following expression:

$$W_t = 4\pi W_0 N \int_0^{\infty} r^2 \exp(-\frac{2r}{r_0} - \frac{4}{3}\pi r^3 N) dr, \quad (6)$$

where N is the activator concentration; r is the distance between centers; r_0 is the radius of the large center.

Theoretical and experimental studies of tunneling from the C-state has revealed that even at an activator ion concentration of In^+ ~ 0.1 mol.% the tunneling probability does not exceed 10^{-2} sec^{-1} [95].

In study [95] tunneling from the A-state was not observed; evidently the probability of A-tunneling is much less than the probability of C-tunneling. Tunneling processes are characterized, as we see from formula (6), by a powerful (exponential) dependence on the distance between the excited activator ions and their concentration. In our case the probability (degree of ionization) within the confidence range (see Figure 13) is independent of the concentration. Consequently the tunnel processes are not responsible for ionization of the mercury-like ions in alkali-halide crystals with a probability of $W_i = 10^6 - 10^7$ sec^{-1}. The independence of our observed ionization processes from the concentration is not the only argument for such a statement. The following experimental facts indicate that tunnel processes do not play a significant role in the ionization of mercury-like ions under intense A-excitation.

1. The induced absorption edge of A^{2+}-centers does not exceed 10^{-8} sec at the same time that tunnel formational processes of the A^{2+}-centers would have a duration comparable to the lifetimes of the excited A-states of the mercury-like ions, i.e., of the order $10^{-6} - 10^{-7}$ sec.

2. The photoconductivity pulse (see § 3) does not exceed the duration of the laser pulse (τ ~ 10^{-8} sec), i.e., the electrons in the conduction band appear simultaneously with excitation. If the A^{2+}- and A^0-centers were to be formed due to tunnel processes, only thermal ionization photoconductivity would be observed with times of the order $10^{-1} - 10^{-4}$ sec and there would be no short photoconductivity pulse.

It would be reasonable to consider tunnel ionization in the light wave field in our discussion of these processes. We know that tunnel ionization of atoms in a static electric field may exist and have been observed experimentally [96]. Tunnel ionization will also occur in an alternating field, particularly a light field if the transition time through the barrier τ_{TYH} is less than the period T of the light oscillations: $\tau_{TYH}/T = \gamma \ll 1$ [96, 97]. The adiabiticity parameter (according to L. V. Keldysh [97]) is expressed by the following formula:

$$\gamma = (2m\omega^2 E_0/e^2E^2)^{1/2}, \qquad (7)$$

where m, e are electron mass and charge; E_o is the ionization potential; E is the light field strength.

Ionization of mercury-like ions in alkali-halide crystals by an UV laser has been reliably observed at intensities of I ~ 10 MW/cm^2 which corresponds to a field strength of E ~ 10^5 V/cm. The value of the parameter γ was estimated for such fields when impurity centers are excited in the alkali-halide crystals in [20] which found a value of $\gamma \sim 800$.

In our case the parameter $\gamma \sim 2000$ for the excited A-state ($E_o \sim 1$ eV) and $\gamma \sim 4000$ for the ground state ($E_o \sim 5$ eV) of the mercury-like ions in alkali-halide crystals.

Therefore the violation of the quasistatic condition of the light field ($\gamma \ll 1$) indicates the impossibility of the realization of the tunnel ionization mechanism in a light field at intensities of I ~ 10 MW/cm^2.

In concluding our discussion of tunnel processes we should note that the low probability of tunneling between the mercury-like ions in alkali-halide crystals ($W_t = 10^{-5} - 10^{-2}$ sec^{-1}) is not an arbitrary occurrence, since the wave functions of their A-states in the alkali-halide crystals are strongly localized and have a low density even at the nuclei of the cations of the second coordination shell [81].

3. Photoionization. Single-quantum photoionization of mercury-like ions in alkali-halide crystals is not possible for excitation to the A-band, since the A-state (relaxed and unrelaxed) is separated from the conduction band by an energy gap of 1-1.5 eV. However the two excitation quanta in the A-band (4-4.5 eV) is sufficient for the electron to make the transition from the ground state of the activator to the continuous spectrum. The two-photon nature of photoionization is also indicated by such experimental facts as the square-law dependence on excitation density of the optical density of induced absorption, the photocurrent and C-luminescence in activated alkali-halide crystals.

In Chapter 1 we demonstrated that the photoionization of mercury-like ions in alkali-halide crystals is related to the excitation of the $^3P_1$-levels, i.e., it occurs from their excited A-states and evidently has a resonant nature. The fact that resonance ionization of the photoexcited centers is observed in experiment is confirmed by the sharp drop in the ionization efficiency with a reduction in temperature of the KI-Tl crystal from room temperature to nitrogen temperature. The absorption spectrum is narrowed and detuning from resonance occurs here, as we see from Figure 14.

Two resonance excitation mechanisms of high-energy states from the absorption of quanta with an energy insufficient for direct excitation exist: The stepwise and cooperative processes [94, 99]. A necessary condition for the effective occurrence of cooperative processes is that the system must contain a level with an energy close to the sum of the energies of the interacting excitation states [94]. In the general case such levels in mercury-like ions are the pseudolocal states related to the high-energy levels of the free electrons. However the independence of our photoionization processes of the activator concentration indicates impossibility of the effective occurrence of cooperative processes [94, 99].

The overwhelming majority of experimental data and an analysis of this data reveal that the ionization of mercury-like ions in alkali-halide crystals under intense excitation in the A-band follow a stepwise mechanism. In this case the intermediate level is the relaxed excited state of the activator.

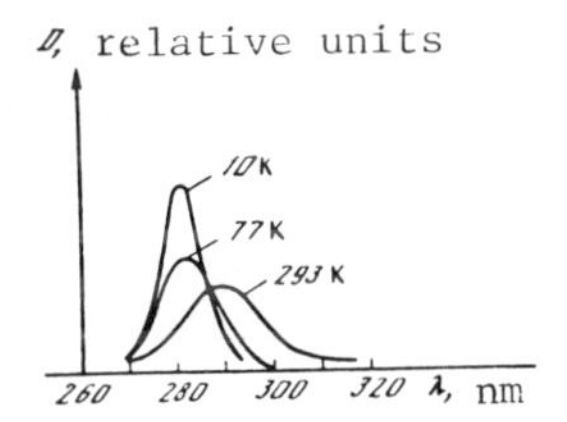

Figure 14. The temperature dependence of the A-absorption band in a KI-Tl crystal.

Apparently there are two reasons for the reduction in the slope of the dependence of the induced absorption on the pump intensity observed at high excitation levels (see Figure 12). First at high intensities ($I \geqslant 3 \cdot 10^7$ W/cm^2) a significant concentration of hole color centers are generated in the crystal, which may increase the probability of electron trapping by the ionized centers and a corresponding reduction in their concentration by the end of laser excitation. Second, the reduction in the degree of nonlinearity may be associated with saturation of the $^1S_0 \rightarrow {}^3P_1$ transition. The criterion for saturation, as we know, is violation of the inequality [96]

$$\int_0^\tau W(t)\,dt \ll 1, \qquad (8)$$

where $W(t)$ is the absorption probability per unit of time; τ is the excitation pulse duration. We will evaluate integral (8) based on the following parameters: $\tau \sim 10^{-8}$ sec; $\sigma \sim 10^{-18}$ cm^2; $I \sim 3 \cdot 10^7$ W/cm^2. The cross-section of the transition $^1S_0 \rightarrow {}^3P_1$ at the center of the band is equal to

$1.3 \cdot 10^{-16}$ cm^2 for the crystal and at the band edge for $\lambda = 308$ nm the cross-section is approximately two orders of magnitude smaller.

With these parameters

$$\int_0^\tau \sigma I(t)dt \sim 0{,}4,$$

i.e., with growth of the excitation intensity ($I \geqslant 3 \cdot 10^7$ W/cm^2) a reduction in the degree of nonlinearity of the stepwise photoionization process is entirely possible.

7. Cross-Section of the Photoionization Processes of Mercury-Like Ions From the Excited States

Therefore intense excitation in the A-absorption band of mercury-like ions in alkali-halide crystals results in, as demonstrated in the preceding sections, stepwise photoionization of the activator and the formation of impurity centers and intrinsic defects in the crystal. Previously (§ 5) we estimated the probability of photoionization by the optical absorption density of Tl^{2+}-centers using Smakula's formula. The photoionization probability is $W_i = 10^6 - 10^7$ sec^{-1} and was independent of the activator concentration.

Measuring the cross-section of the optical transition from the excited state of the activator to the conduction band of the crystal is of interest. The phototransition cross-section - one of the most important characteristics of optical spectroscopy - is a parameter that objectively characterizes the optical transition. Measurements of the cross-section are not of purely scientific interest. Knowledge of the cross-section makes it possible to estimate the efficiency of defect generation in activated alkali-halide crystals which is important from the viewpoint of selecting media that are promising for optical data recording and for developing lasers operating by the defects in alkali-halide crystals [8, 100].

Theoretical estimates of the cross-sections of photoionization transitions are quite complex due to possible distortions to the wave functions of the band states in the defect region [17] and hence it is necessary to experimentally measure the photoionization cross-section. We may obtain the power of the cross-section from available estimates of the photoionization probability by the familiar relation $W_i = I\sigma_i$, where I is the excitation intensity and σ_i is the photoionization cross-section. A value of $W_i = 10^6 - 10^7$ sec^{-1} at an intensity of $I \sim 10^{24}$ photons/($cm^2 \cdot$ sec) corresponds to cross-sections of $\sigma_i = 10^{-17} - 10^{-18}$ cm^2.

The absorption method which has been used to estimate the probability has, in our view, certain drawbacks that make it impossible to measure the cross-section with sufficient accuracy. Such drawbacks include the need to use high excitation intensities ($I \geqslant 10$ MW/cm^2) which may result in possible optical breakdown of the defects, together with a narrow dynamic range and overlap of the absorption bands of the defects created.

There exists a method for determining the photoionization cross-sections of an activator by the photoresponse of the crystals [83]. In spite of the fact that we were evidently successful in recording and identifying photoelectrons whose source was activator ions, the present study did not employ the photoelectric method for measuring the cross-sections. This is due to such factors as the possible competition of two-stage and two-photon processes in the activated alkali-halide crystals and the need for an exact measurement of a large number of parameters.

In this study we employed the luminescence method of measuring cross-sections whose general description is given below. The quantum efficiency of luminescence of mercury-like ions in alkali-halide crystals under excitation by regular light sources (lamp or spark excitation) is independent of the excitation intensity and is close to unity [101]. In the case of intense laser excitation where photoionization of the activator from the excited state is observed, the quantum efficiency of intracenter luminescence will drop, since some of the electrons from the excited state are expanded in the formation of the electron defects: The Tl^{o}- and F-centers.

Therefore by measuring the dependence of the relative quantum efficiency of intracenter luminescence on the excitation intensity we may determine the portion of ionized centers from the total number of excited centers (the degree of ionization) and measure the photoionization cross-section of the excited state of the activator. The advantages of this method include the fact that it is necessary to measure only one parameter: The amplitude of the luminescence signal and auxiliary measurements (calibration of the light filters) are performed only once with good accuracy (5-10%).

We will examine in greater detail the two-step photoionization model we employed for quantitative calculations (Figure 15). The quantum absorption and emission processes by the mercury-like ions in the alkali-halide crystals may be represented by a four-level scheme.

The relaxed excited state of the activator may have a complex energy and spatial structure [54-58] with a thermal equilibrium distribution existing between its components. Since both the radiative and metastable levels are populated under excitation (T = 300 K) [9], then ionization will most likely occur from both the radiative level $^3P_1$ as well as the metastable level $^3P_0$.

For our quantitative calculations we assumed that by absorbing a light quantum the electron will make the transition to the excited level of the luminescence center and will make the transition to the conduction band by the second quantum and will then be localized at some trap center, i.e., the electron will not return to the ionized center. In this model we ignore changes in the population of the relaxed excited state due to spontaneous and stimulated emissions and we also ignore the reverse electron trapping by ionized centers.

Changes in the population of the relaxed excited state due to spontaneous emission is not significant, since the excitation duration (10^{-8} sec) is much less than the lifetime of the excited states of the mercury-like ions at room temperature (10^{-6} - 10^{-7} sec) [15, 16]. An estimate of the probability of induced emission $W_{ИНД}$ has demonstrated that at the excitation intensity achieved in our experiment of $I \sim 10^{25}$ photons · cm^{-2} · sec^{-1}, $W_{ИНД} = 10^5 - 10^6$ sec^{-1}, i.e., it does not exceed the probability of spontaneous emission of A-luminescence of mercury-like ions in alkali-halide crystals. The influence of repeated trapping of the band electron by the ionized center will be discussed below.

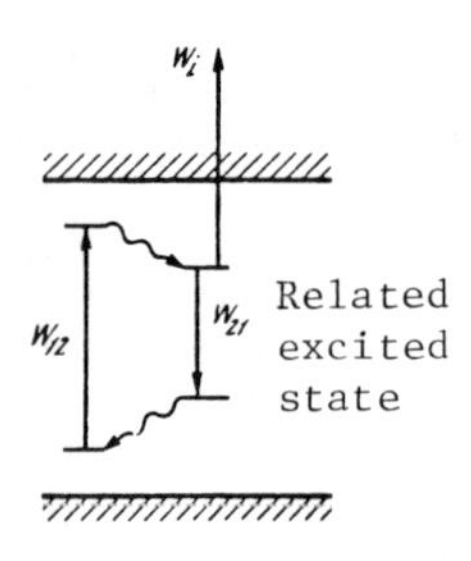

Figure 15. Two-photon stepwise photoionization scheme of mercury-like ions in alkali-halide crystals.

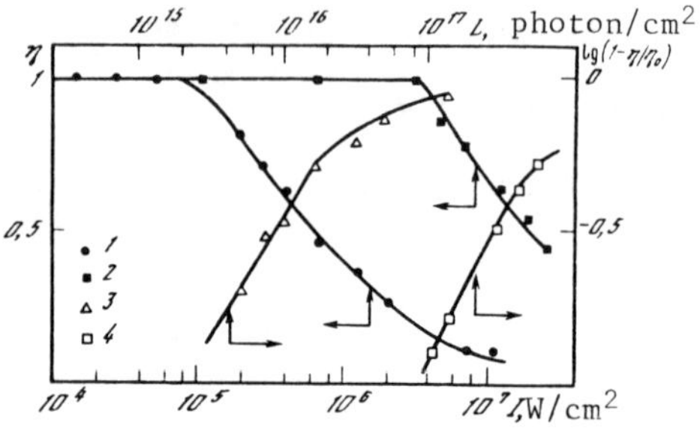

Figure 16. The relative quantum efficiency of ionization plotted against the excitation intensity.

1,3 - KI-Tl; 2, 4 - KI-Tl.

The arrows by the curves indicate their coordinate axes.

Subject to these assumptions the equations for the population on the ground (n_1) and excited (n_2) levels may be written in the following manner:

$$dn_1/dt = -\sigma_{12} I n_1, \tag{9}$$

$$dn_2/dt = \sigma_{12} I n_1 - \sigma_i I n_2. \tag{10}$$

We may determine the population n_2 at the excited level after the excitation pulse from equations (9) and (10)

$$n_2 = N \frac{\sigma_{12}}{\sigma_{12} - \sigma_i} [\exp(-\sigma_i L) - \exp(-\sigma_{12} L)], \tag{11}$$

where $L = \int_0 I(t)dt;\ N$; N is the activator concentration. The quantity of observed quanta n_2' after the excitation pulse is determined from equation (9):

$$n_2' = N[1 - \exp(-\sigma_{12}L)]. \tag{12}$$

The quantum efficiency of luminescence η (in relative units) is the ratio of the number of emitted quanta n_2 to the number of observed quanta n_2':

$$\eta = \frac{n_2}{n_2'} = \frac{\sigma_{12}[\exp(-\sigma_i L) - \exp(-\sigma_{12}L)]}{(\sigma_{12} - \sigma_i)[1 - \exp(-\sigma_{12}L)]}. \tag{13}$$

Therefore if we know the value of the quantum efficiency of luminescence for some given value of L we may determine the photoionization cross-section. In practice the relative quantum efficiency $\eta = \eta(L)/\eta_o$ was measured, where η_o is the relative quantum efficiency of luminescence under weak excitation, i.e., it was normalized to unity. With weak excitation (small values of L) the equations for the population on the excited level and the number of absorbed quanta may be simplified:

$$n_2 = N\sigma_{12}L[1 - L(\sigma_{12} + \sigma_i)/2], \tag{14}$$

$$n_2' = N\sigma_{12}L(1 - \sigma_{12}L/2). \tag{15}$$

In this case

$$\eta = n_2/n_2' \approx 1 - \sigma_i L/2, \tag{16}$$

or

$$1 - \eta \approx \sigma_i L/2. \tag{17}$$

Expression (17) is nothing other than the quantum efficiency of ionization η_i and may be used to estimate the photoionization cross-section [29].

We will now examine the experimental results. Figure 16 gives the experimental values of the relative quantum efficiency of luminescence η and the quantum efficiency of photoionization $(1 - \eta)$ as a function of the excitation intensity for KI-Tl and KI-In crystals. As we see from the figure there exists for both crystals a range of excitation intensity where η remains constant, and with a further increase in intensity it decays, i.e., "ionization quenching" is observed. We may determine the photoionization cross-section σ_i by the relation $(1 - \eta) = f(L)$ (Figure 16, curves 3 and 4). We will give the values of the photoionization cross-sections from the excited states of the In^+ and Tl^+ ions in KI (σ_{12}, taken from [9] and σ_i determined by formula (17) from Figure 16): Or a KI-Tl crystal ($\sigma_{12} = 13 \cdot 10^{-17}$ cm^2, $\sigma_i = 3 \cdot 10^{-17}$ cm^2; for a KI-In crystal $\sigma_{12} = 0.59 \cdot 10^{-17}$ cm^2, $\sigma_i = 4 \cdot 10^{-17}$ cm^2.) The luminescence measurements are a good supplement to the absorption measurements, since it is possible to detect photoionization by means of luminescence at much lower intensities than by the absorption of the A^{2+}- and A^{o}-centers. In the interval I = = 10^5 - 10^6 W/cm^2 the quantum efficiency of ionization ($1 - \eta$) is linearally dependent on L (or I) (see Figure 16); since ($1 - \eta$) is the ratio of the number of ionized centers to the excitation intensity, the experimentally-observed relation is

$$1 - \eta = AI, \tag{18}$$

where A is a constant and is nothing other than the quadratic dependence of the number of ionized centers N_i on the excitation intensity:

$$N_i = AI^2. \tag{19}$$

Relation (19) is one additional proof of the stepwise ionization of the activator.

Calculations using formulae (14)-(17) are simple although their range of application is limited due to low excitation intensities. Therefore in order to achieve a comprehensive physical picture of these phenomena we will examine the experimental dependence (curve 1) of the amplitude of the luminescence pulse of a KI-In crystal on the quantity L shown in Figure 17 together with calculated curves based on equations (11) and (12). Curve 2 derived from equation (12) corresponds to the case where there is no ionization. The knee in curve 2 corresponds to the onset of saturation of the $^1S_o \rightarrow {}^3P_1$ transition. By taking the ratio of the amplitude of the luminescence signal n_2 to the number of absorbed quantum n_2' with a fixed L, we may calculate the ionization cross-section σ_i and plot the relation n_2 = = $f(L)$ (Figure 18, curve 3).

Our method for calculating the photoionization cross-sections makes it possible to describe the experimental curve up to values of L = 5 · 10^{16} photons · cm^{-2} using equation (11), i.e., to values of L where the luminescence signal is reduced by a factor of 1.5-2 due to ionization. We attribute the deviation in the experimental curve from the theoretical curve values of L $\geqslant$ 5 · 10^{16} photons/cm^2 to an increase in the probability of repeat trapping by the ionized centers due to an increase in their concentration. We know that the electron recombination cross-section at the ionized Tl^{2+} center in KCl is more than two orders of magnitude greater than the trap cross-section at the Tl^{+} ion. Evidently an analogous relation holds for other crystals as well, since the recombination and trap cross-sections are determined primarily by the charge state of the centers [102]. Hence in order to determine the photoionization cross-sections we employed experimental data with small values of L $\leqslant$ 10^{16} photons/cm^2 where the percentage of ionized centers is low (less than 0.01).

Figure 18 gives the experimental relations for the luminescence pulse amplitudes for KCl-In, KI-Tl, KBr-In crystals. These relations were used to calculate the photoionization cross-sections σ_i provided in Table 1 (σ_{12} is the absorption cross-section at the center of the A-band). The absorption cross-section values at the center of the A-band are taken from [9] or are determined by Smakula's formula. For all crystals except the KI-Tl crystal the photoionization cross-section from the lower excited state is independent of the concentrations C.

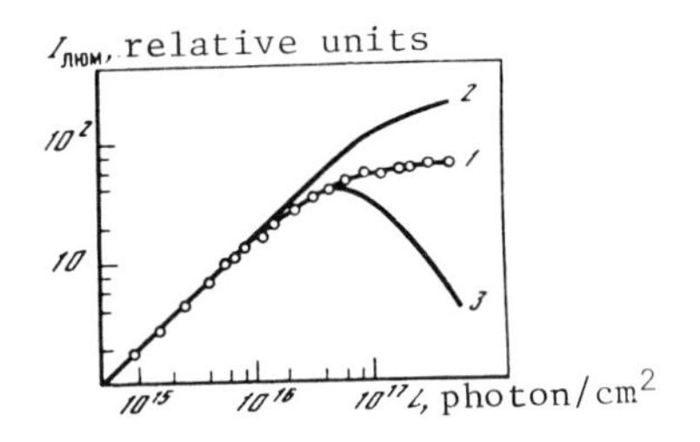

Figure 17. Experimental (1) and calculated (2, 3) dependencies of the luminescence amplitude on the quantity of quanta absorbed over the excitation period (KI-In crystal).

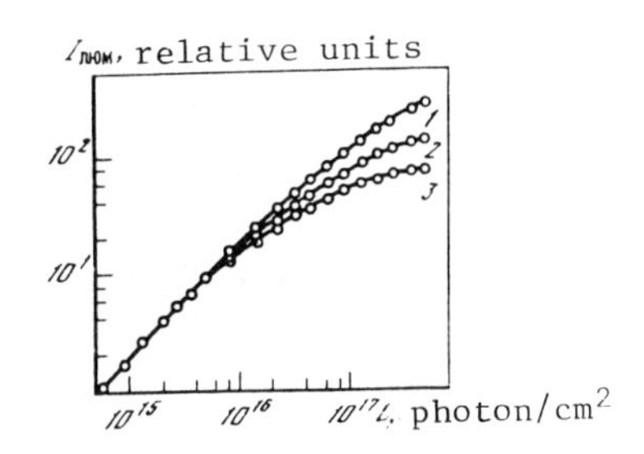

Figure 18. Experimental dependencies of the luminescence amplitude on the quantity of quanta absorbed over the excitation period for KCl-In (1), KT-Tl (2) and KBr-In (3) crystals.

Since the KI-Tl crystals were pumped (λ = 308 nm) to the longwave edge of the A-band, the increase in the ionization cross-section with an increase in the activator concentration may be attributed to a new absorption band that arises due to the formation of paired $(Tl^+)_2$ centers [103].

We should note that we were not able to determine the concentration dependence of the ionization probability for KI-Tl using absorption techniques, which possibly indicates the low information content of the absorption method requiring very high ($I \geqslant 20$ MW/cm^2) excitation intensities.

The primary sources of error in determining the photoionization cross-sections include measurement of the laser emission intensity and accounting for its spatial and temporal structure. We accounted for the time dependence of the intensity by using an oscillogram of the laser pulse for determining the quantity L. We carried out a test of the intensity distribution $I(x, y)$ in the plane perpendicular to the beam axis by the burn-in of film from a number of laser pulses with a variable degree of energy damping. It is established that the $I(x)$ and $I(y)$ distributions may be represented by smooth curves with a maximum at the center of the beam.

We should note that in the general case we should introduce volumetric integration into formulae (11) and (12) for the volume over which luminescence is recorded. In the experiment we projected onto the monochromator slit only the portion of the crystal bulk in which the excitation intensity varied due to absorption to an insignificant degree (~ 10-15%). This formulation of the experiment made it possible to avoid volume integration and to simplify the calculations using formulae (11) and (12). Overall the measurement accuracy for measuring the absolute cross-sections was 50-70% with a relative measurement accuracy of 20%.

Table 1

Photoionization cross-sections from the excited states of mercury-like ions in alkali-halide crystals

Crystal	C, 10^{17} cm^{-3}	σ_{12}, 10^{-17} cm^2	σ_i, 10^{-17} cm^2
KI-Tl	10	13	3
KI-Tl	1	13	1
KCl-In	1-100	0.2	1.6
KBr-In	1-100	0.5	4.5
Kl-In	1-100	0.6	4
KBr-Sn	1-100	0.7	4.5

It is clear from Table 1 that the values of the photoionization cross-sections of mercury-like ions in alkali-halide crystals were equal to $(1\text{-}4) \cdot 10^{-17}$ cm^2 which, for all test crystals with the exception of KI-Tl exceeds the absorption cross-section at the center of the A-band. Our derived values for the cross-sections were close to the cross-sections of the allowed electron-dipole transitions. This fact may be a serious obstacle to obtaining stimulated emission from mercury-like ions in alkali-halide crystals.

The following situation may serve as a criterion for evaluating the promise of building lasers employing mercury-like ions in alkali-halide crystals: The quantum pump energy cannot exceed the energy gap between the upper working level and the conduction band. If we account for the existence in the relaxed excited state of the $^3P_0$ metastable level and the phenomenon of induced absorption of defects formed from the pump process [21-24] then it is not likely that lasers will be built to operate by the Tl^+, In^+, Sn^{2+} mercury-like ions in alkali-halide crystals. Moreover it is still possible to obtain lasing in the IR from defects including mercury-like ions [8].

One interesting problem that logically follows from these investigations is the following: Does ionization occur from the transitions to the continuous spectrum or to the quasilocal level of the activator in the conduction band of the alkali-halide crystals [32, 104]?

In order to test this notion we investigated the dependencies of the relative quantum efficiency of A-luminescence on the excitation intensity in the A- and C-absorption bands of In^+ ions in KBr and KI and the Sn^{2+} ions in KBr [44]. An excimer XeCl laser (λ = 308 nm) and the fourth harmonic from an Nd:YAG laser (λ = 266 nm) were used as the excitation source for the A- and C-bands, respectively. The experiments were carried out at room temperature.

Figure 19 gives the relations for the relative quantum efficiency of A-luminescence for excitation to the $^3P_1$- and $^1P_1$-states.

It is clear that for all test crystals the reduction in the relative quantum efficiency of luminescence with C-excitation occurs at intensities significantly greater than in A-excitation. This correlates well with the onset of induced absorption of defects resulting from activator ionization. In order to understand the processes that occur from intense excitation in the A- and C-bands we will consider an arbitrary energy scheme of mercury-like ions in alkali-halide crystals (Figure 20).

We know that at room temperature the emission spectrum of mercury-like ions is identical for both A- and C-excitation and consists primarily of a single-A-luminescence band where the quantum efficiency of emission is close to unity in both cases [101]. Consequently relaxation to the $^3P_1$-level occurs for C-excitation, and even at helium temperatures the excitation relaxation time from the $^1P_1$- to the $^3P_1$-state is less than 10^{-9} sec [105]. Consequently for both A- and C-excitation virtually the higher population is on the $^3P_1$-level from which spontaneous emission and (in our case) ionization originates.

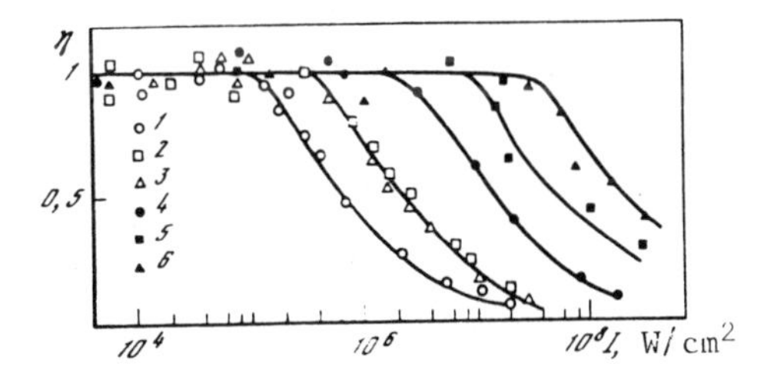

Figure 19. The relative quantum efficiency of A-luminescence of mercury-like ions in alkali-halide crystals plotted as a function of the excitation intensity in the A-band (1-3) and in the C-band (4-6).

1 - KCl-In; 2, 5 - KBr-In; 6 - KBr-Sn; 4 - KI-In.

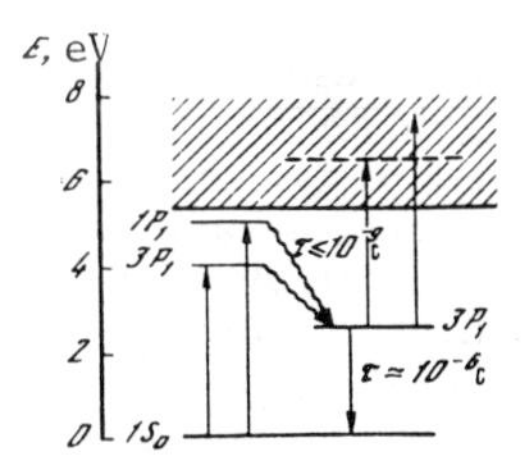

Figure 20. Relative energy diagram of the levels of mercury-like ions in alkali-halide crystals.

Since a drop in the relative quantum efficiency of luminescence is observed in the excitation to the $^1P_1$-state at significantly higher pump densities compared to the excitation to the $^3P_1$-level, and ionization takes place from this same level we may assume that the cause of ionization is not luminescence reabsorption [104], but rather transitions from the $^3P_1$-level under the action of excitation quanta.

Moreover, since the quantum efficiency of ionization is dependent on the excitation wavelength the transition from the $^3P_1$-level under A-excitation (4 eV) is a resonance transition and evidently goes to the quasilocal activator level rather than the continuous spectrum. This is indicated by the different excitation intensity thresholds for observing ionization of the different ions in the different matrices.

Two spectral points are insufficient to establish the nature of the optical transition from the excited state, i.e., it is necessary to expand the spectral range of the investigations. The results of such research may be found in Chapter 3.

CHAPTER 3

OPTICAL TRANSITIONS FROM THE EXCITED STATE OF MERCURY-LIKE IONS TO THE ALKALI-HALIDE CRYSTAL CONTINUUM

Several absorption mechansims have been observed in activated crystals: Intrinsic absorption or fundamental absorption; impurity absorption; exciton absorption; free carrier absorption; intraband absorption, etc. We will be interested in impurity absorption, i.e., transitions from the valence band to the impurity levels or, closer to our case, electron transitions from the impurity centers to the conduction band.

The present chapter is devoted to the search and investigation of absorption from the excited state of activator centers to the band states.

Quasilocal activator levels and their influence on the photoionization processes of luminescence centers are discussed. Hence the beginning of the chapter provides the fundamental theoretical aspects and certain experimental results relating to this problem.

8. Phototransitions Between the Local and Band States in Crystals

The overwhelming majority of theoretical and experimental research on the transitions between local and band states have been carried out with application to semiconductor physics. This is not surprising since the presence of impurities having impurity levels in the bandgap may produce many useful and undesirable effects in semiconductors and may both expand and limit the wide variety of semiconductor applications [106-110]. There are few studies of such transitions for direct application to broadband dielectrics and these primarily relate to the photoionization of various types of F-centers and other defects [14, 17, 35, 66, 111].

From the viewpoint of the fundamental regularities of optical transitions in solids the division between semiconductors and dielectrics becomes meaningless, since the differential is primarily based on quantitative concepts [109]. Therefore the majority of fundamental results in semiconductor physics are also qualitatively applicable to a discussion of our photoionization processes from the excited state of mercury-like ions in alkali-halide crystals. The closest analogies are evidently to the ionization processes of the deep impurity levels in semiconductors, since the $^3P_1$-state of mercury-like ions in alkali-halide crystals is, as a rule, 1.5 eV below the bottom of the conduction band.

Optical transitions in solids have special features compared to analogous transitions in free atoms and ions; these are related to electron-phonon interaction and the specific selection rules. Therefore the following selection rule must be observed for direct transitions:

$$\mathbf{k}_2 = \mathbf{k}_1 + \mathbf{g}, \tag{20}$$

where k_2 and k_1 are the quasiwave electron vectors before and after the transition; **g** is the wave vector of the photon.

We know that for the visible and near UV the photon momentum is negligible compared to the quasimomentum of a thermal electron [107], and hence selection rule (20) adopts a simpler form:

$$\mathbf{k}_2 \approx \mathbf{k}_1. \tag{21}$$

We will not consider indirect transitions, i.e., transitions involving photons and phonons simultaneously, since their probability is two to three orders of magnitude less than the probability of direct transitions [110]; in our case the photoionization cross-sections are close to the cross-sections of the allowed transitions.

We will consider the more general photoionization regularities of impurities with the following simplifications [66]:

1) The lattice state does not change in the transition;

2) the state of the free carrier is one of the unperturbed states of the conduction band;

3) the conduction band is isotropic, nondegenerate and parabolic:

$$E_k = \hbar^2 k^2 / 2m. \tag{22}$$

In this case the shape of the absorption line is as follows [66]:

$$G(\omega) \sim m^* |A_{12}|^2 \left(\frac{2m^*}{\hbar^2}\right)^{1/2} \frac{(\hbar\omega - E_1)^{1/2}}{\hbar\omega}, \tag{23}$$

where A_{12} is the transition matrix element; E_1 is the threshold ionization energy.

It follows from expression (23) that the photoionization cross-section is proportional to the state density in the conduction band and may be dependent on the behavior of the matrix element. We will use the wave function of the localized state as in [107] in the form of the s-state function of a single-electron atomic system:

$$\Psi_1 = Ce^{-\varkappa\rho}e^{-\frac{i}{\hbar}E_1 t}. \tag{24}$$

The quantity $\varkappa^{-1}/2$ is equal to the distance over which the probability of finding the electron is reduced by a factor of e.

The calculations carried out in [107] reveal that with the original wave function (24) the probability of electron transition to the final state of the free carrier in the unperturbed conduction band may be represented as

$$W_{12} \sim \frac{\varkappa^5 k_2^2}{[\varkappa^2 + (k_2 - g)^2]^4} I, \tag{25}$$

where I is the photon flux.

Neglecting in (25) the wave vector of the photon g, we will consider the quantity

$$\beta = \varkappa^5 k_2^2/(\varkappa^2 + k_2^2)^4. \tag{26}$$

If we consider W_{12} as a function of k_2^2, W_{12} then passes through a maximum whose position may be found from the condition

$$d\beta/dk_2^2 = 0, \tag{27}$$

or

$$k_2 = \varkappa/\sqrt{3}. \tag{28}$$

Due to the sharp reduction in W_{12} with large k_2 ($W_{12} \sim k_2^{-6}$ when $k_2 \gg \varkappa$) we may take the quasiwave vector of the electron to vary in the interval $0 < k_2 < \varkappa$, i.e., as the light is absorbed by electrons in the localized states those photons that convert the electrons to a state with an energy $E_2(k_2) = E_1 + \hbar\omega$ are preferentially absorbed, so

$$k_2 \approx \varkappa. \tag{29}$$

Equation (29) is a very important selection rule in the photoionization of localized centers in crystals. This equation results from the uncertainty relation [107]. The uncertainty of the momentum of an electron localized in a potential well of dimensions $\varkappa^{-1}$ is equal to

$$\Delta p \sim \hbar/\varkappa^{-1} = \hbar\varkappa. \qquad (30)$$

In the transition to a free state the electron momentum $\hbar k_2 \approx \Delta p$, i.e., $k_2 \sim \varkappa$ in complete agreement with (29). Derived relations (25) and (29) reveal that the absorption spectrum from local states to continuum (22) will take the form of possibly a broad yet limited band. This is because the electron in the localized state occupies a limited region of the quasi-momentum space.

Analogous results for photoionization of impurities have also been obtained in the familiar surveys by Kohn [112] and Johnson [113]. The primary difficulty of the theory of optical transitions between impurity and band states is in searching wave functions describing the motion of carriers localized near impurity centers [110]. Minor defects ($E_1 \leqslant 0.1$ eV) are satisfactorily described by the hydrogen-like model and the Coulomb potential may be used for their calculation [110].

Searching the wave functions of the deep centers is a more complex problem then for the shallow donors and acceptors and the theory of optical transitions involving deep centers has only begun to be developed [110]. Calculations often employ the approach developed by G. Lucovsky [114] where the potential of the defect is approximated by a delta function with a deep well equal to the binding energy. A. Milnes has noted [106] that the applicability of such a method is limited to the case of very deep impurity levels. In the Lucovsky model the photoionization cross-section $\sigma(\hbar\omega) = 0$ when $\hbar\omega = E_1$, and with growth of $\hbar\omega$ it increases rapidly, reaching a maximum for $\hbar\omega = 2E_1$ and then decays as $(\hbar\omega)^{-3/2}$ when $\hbar\omega \gg E_1$. The derived results are in satisfactory agreement with experiment for an indium impurity in silicon ($E_1 = 0.15$ eV).

In actual crystals the conduction band may be more complex than (22) and the electron state density may be influenced by the impurities of other bands as well as the quasilocal levels of the impurity center. Thus if the conduction band consists of several subbands, in the photoionization of the impurity the electrons may be pushed up to the upper subbands. One example is direct transitions from the donor levels to the higher-lying conduction band in gallium phosphide doped by selium, silicon and tellurium [115].

The quasilocal levels of the activator may have a perturbing influence on the conduction band; in this case both the state density and the matrix element of the transition change. We should note that the majority of theoretical and experimental research on the photoionization of impurity centers in solids have been carried out ignoring the perturbation of the continuum states near the energy of the pseudolocal states. At the same time the cases where the excited states of the impurity absorbing system are accumulated on the energy scale at the continuum states of the crystal-matrix are evidently rare. An often-cited example includes the L-absorption bands of the F-centers [14], the C-bands in CaS-Pb [116], and certain Tl^{0} transitions in KCl [36]. There are also a number of examples in semiconductor crystals [37-39].

The pseudolocal states in activated alkali-halide crystals have been investigated by Ch. B. Lushchik, N. N. Kristofel, G. S. Zavt et al. [1, 12, 17, 31, 116]. Ch. B. Lushchik applied the semiempirical approach to calculating the position of the quasilocal levels on the energy scale based on the compression coefficients of the levels of a free ion in the crystalline lattice field [1, 12]. N. N. Kristofel and G. S. Zavt achieved a number of fundamental theoretical results; specifically, the calculation of the spectra [31]. The pseudolocal state, as noted in [31] must be well observed if the excited level of the impurity has symmetry absent in the band. Moreover we note that in the general case the spectrum of impurity photoabsorption has a composite form that reflects both the distribution of the density of states in an ideal lattice and the actual nature of the band states.

Page et al. [111] have investigated the quasilocal states of the F-centers. The calculation of the absorption cross-sections from the ground level is in satisfactory agreement with the experimental data; the peaks are assigned to the matrix element immediately above the threshold, at the same time that at high energies the peaks are related to the structure of the electron state density. Study [35] is another example of satisfactory agreement between theory and experiment; this study obtained the following dependence of the photoionization cross-section of the F'-center on the photon energy:

$$\sigma(E) \sim [(E - E_1)/E_1]^{3/2}, \tag{31}$$

where E_1 is the threshold energy.

The generation of quasilocal states in solids has an analogy to such a phenomenon as the self-ionization state in free atoms [33, 34]. A consistent theory of self-ionized resonances was first formulated by Fano [33]. According to this theory interaction between the discrete level and the continuum serves to broaden the local level to the band whose density of states is determined by a Lorentz curve with half-width $\pi|V_E|^2$, where V_E is the matrix element of their hybridization. The structure of the resonance is asymmetric due to the interference of the matrix elements of the transitions, with the increase in amplitude, on the one hand, and with a decrease in amplitude, on the other. Without interference the discrete level in the continuum appears as a symmetric absorption line against a background of a smooth continuous spectrum. Interference violates simple summation of the contributions and produces noticeable asymmetry in the majority of self-ionization lines. Let $\Gamma/2$ be the half-width of the resonance state, and E_0 be the resonance energy. We will measure the energy in units of

$$\varepsilon = (E - E_0) / \tfrac{1}{2}\Gamma, \tag{32}$$

while q is the relative amplitude of the oscillator force of the resonance state compared to the scattering states. Then, according to Fano, the line shape near resonance is determined by the expression

$$\alpha = 1 + \frac{q^2 - 1}{1 + \varepsilon^2} + \frac{2q\varepsilon}{1 + \varepsilon^2}. \tag{33}$$

The terms in the right half of (33) are background absorption, the symmetric Lorentz peak and the result of interaction between the resonance and scattering transitions. Antiresonance exists when $\varepsilon = -q$. The curves deriving from (33) are shown in Figure 21.

The resonance interaction with the continuum states was investigated for the first time in inert gas atoms. Examples of asymmetric resonances in solids as well may also be found: Exciton resonances in Cu_2O [117] and transitions in $KMgF_3:V^{2+}$ [66]. The self-ionization lines in actual systems may have a different shape from the calculated curves found in Figure 21. Thus there is no minimum for certain self-ionization transitions in Xe, and the line contour is a Lorentz contour [34]. In solids the interaction picture is even more complex due to electron-phonon interaction: The influence of phonon broadening serves to reduce antiresonance [118].

Thus the spectral dependence of the photoionization cross-section of impurity centers in solids is determined by many factors and establishing the actual physical picture in each specific case is a complex experimental and theoretical task: There are, evidently, no universal criteria here.

The concepts developed by Fano [33] have not been used extensively to investigate quasilocal levels in solids. There are exceptions which include [39] which investigates the properties of the pseudolocal states in GaAs:Cr. Among the experimental studies on nonlinear spectroscopy we cannot overlook the excellent study by Mollenauer [119] devoted to an investigation of the high-excitation states of the F_2^+-centers in KF. Mollenauer used the polarization technique to identify absorption from the lower excited state of the F_2^+-center to the upper-lying states despite the overlap with absorption from the ground state. The experimental results are in good agreement with calculations of the F_2^+-center using the H_2^+-model.

In conclusion we again emphasize that the absorption of the activator from the local states to the band states is influenced by many factors whose degree of influence in the absorption mechanism must be accounted for in each specific case. Moreover there are fundamental regularities related, specifically, to the quasiwave vector selection rule (29).

9. Absorption From the Excited State of Mercury-Like Ions in Alkali-Halide Crystals

An investigation of the quantum efficiency from excitation in the A- and C-bands has revealed that the transition from the $^3P_1$-state to the band state in the 4 eV range evidently has a resonance nature. Investigating the spectral dependence of the phototransitions from the excited state is of indisputable interest, although we considered the method analogous to measuring the cross-sections by the quantum efficiency to be inefficient in this case due to the need for multiple measurements at various points in the spectrum.

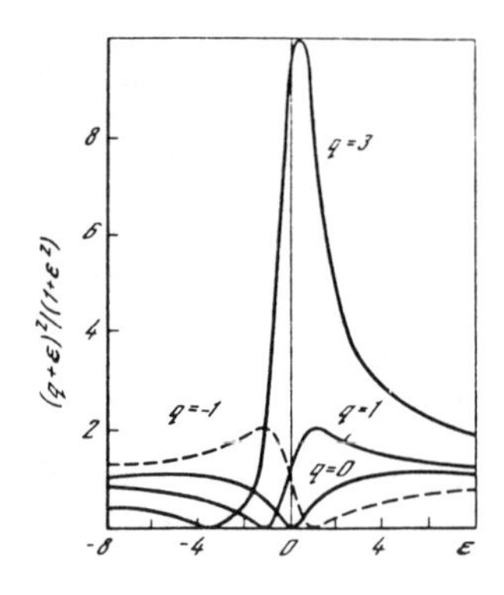

Figure 21. Self-ionization resonance line contours.

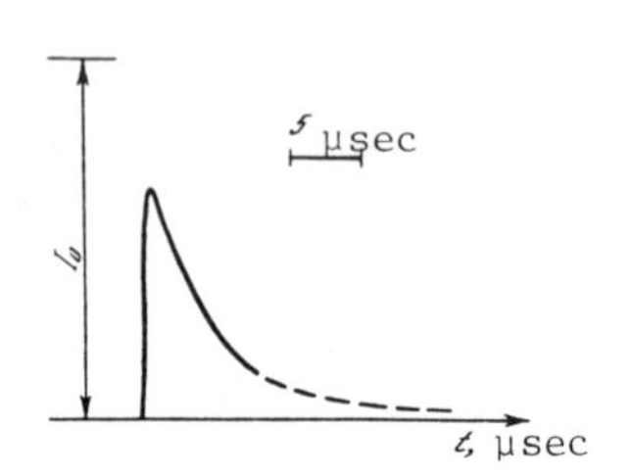

Figure 22. The kinetics of absorption from the excited state of In^+ ions in KBr.

The dotted line curve represents the temperature-dependent part.

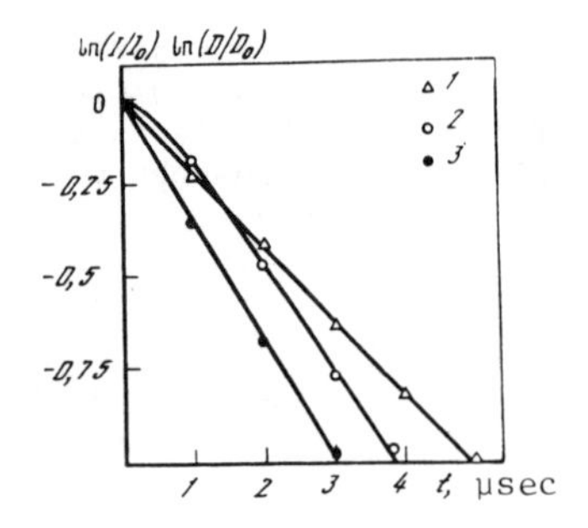

Figure 23. The luminescence damping kinetics (solid line curves) and the absorption kinetics (markers).

1 - KCl-In (T = 100-293 K); 2 - KBr-In (T = 100 K); 3 - KBr-In (T = 293 K).

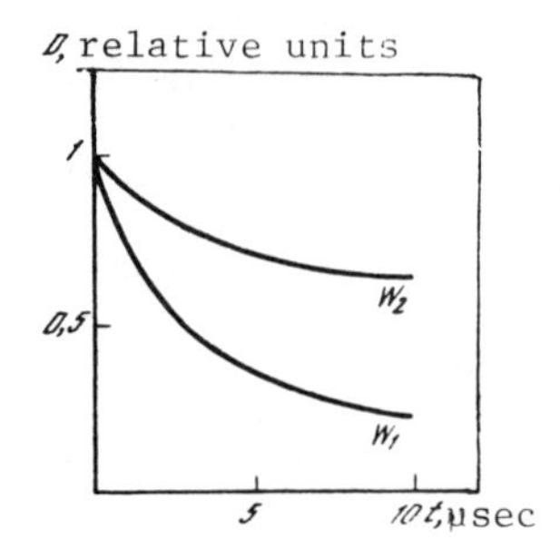

Figure 24. Absorption kinetics according to formula (37) for $W_1 = 3 \cdot 10^5\ sec^{-1}$ and $W_2 = 10^5\ sec^{-1}$.

We attempted to detect optical absorption from the excited A-state of mercury-like ions in alkali-halide crystals. It is obvious that in order to register absorption from the excited state it is necessary to maintain a high population on the excited level. To achieve this in the continuous mode as done by Mollenauer in his investigation of the high-energy levels of the F_2^+-centers in KF [119] was not possible in our case due to a lack of CW lasers operating in the 4 eV range. Hence in our search for absorption from the excited state of mercury-like ions in alkali-halide crystals we used pulsed excitation, i.e., we employed the technique analogous to that used to investigate the induced absorption of activator defects (see Chapter 1). In the final analysis the experiment was reduced to using two-step spectroscopy of the excited states.

We selected indium phosphors (KI-In, KBr-In, KCl-In) as our test samples, since XeCl laser emission correlates well with their A-absorption bands and, moreover, data on the cross-sections of the phototransitions from the excited states were already available for these phosphors. By careful tuning of the experimental set-up (see Figure 1) we were able to record absorption whose damping duration was close to the spontaneous decay time of the lower excited state of the indium ions in the alkali-halide crystals. The absorption signals were observed primarily in the 3-4 eV spectral range; a typical absorption pulse in a KBr-In crystal is shown in Figure 22. The method used to process the absorption signals in order to obtain the temporal variation in optical density was analogous to that described previously in Chapter 1.

As we know under intense excitation in the A-band of mercury-like ions in alkali-halide crystals a variety of impurity centers and intrinsic defects are formed [21-24] having broad absorption bands. The observed absorption could be attributed not only to absorption from the excited state but also to a change in the concentration of defects due to their thermal ionization and photoionization as well as tunnel recombination.

In order to establish the nature of the absorption detected we carried out a series of experiments: The dependencies of the absorption decay time on temperature, excitation intensity and activator concentration were investigated. Now we will consider possible physical phenomena that could result in the observed effects and will analyze experimental results.

1. Thermal ionization of defects. This mechanism was already analyzed in Chapter 2 which demonstrated that the decay time of the induced absorption amounts to hundreds of microseconds and is determined by the thermal ionization of the electron color centers A^0. Moreover it is also possible that "shallow traps" are formed whose thermal decay may occur over shorter time periods than the thermal ionization of the A^0-centers and hence it is necessary to bear in mind this mechanism as well.

Temperature investigations in the T = 100-600 K range have demonstrated that our absorption pulse consists of two parts: A temperature-dependent part and a temperature-independent part (see Figure 22). The temperature-dependent part has a duration of the order of hundreds of microseconds and is attributable to the defects. Subsequently it was determined that the temperature-independent part was not in reality temperature-independent in the strictest sense and does experience small changes in the 100-300 K range. We know that the lifetimes (the slow component) of the $^3P_1$-states of the In^+ ions in KI and KBr also experience small variations in this range [16]. A special test has revealed that the change in the damping time of the detected absorption exactly coincides with the temperature changes in the lifetimes of the excited states of the activator.

Figure 23 gives the absorption and luminescence damping curves for In^+ ions in KBr and KCl at T = 100 and 293 K. Since the probability of thermal ionization (according to formula (2)) has an exponential temperature dependence, the observed absorption cannot be attributed to thermal ionization of the shallow traps and defects.

2. Photoionization of defects. The In^0 electron color centers have broad absorption bands and light irradiation in these bands may ionize the defects [73]. Such irradiation may be achieved by luminescence photons of the activator ions and in the general case may produce absorption signals of the probe beam whose duration is comparable to the duration of luminescence damping.

We will analyze the possibility of such a process. Let the luminescence follow the regular law

$$I = I_0 \exp(-t/\tau), \tag{34}$$

where τ is the lifetime of the excited state. The number of nonionized defects per unit of time in the ionization by luminescence photons will be expressed by the following equation:

$$dn/dt = -n\sigma_i I(t), \tag{35}$$

where σ_i is the ionization cross-section. Substituting (34) into (35) we obtain the differential equation

$$dn/dt = -n\sigma_i I_0 \exp(-t/\tau). \tag{36}$$

The solution of equation (36) is:

$$n = n_0 \exp[-\sigma_i I_0 \tau (1 - e^{-t/\tau})]. \tag{37}$$

It follows from (37) that the duration of absorption of the probe beam by the defects during their photoionization by the luminescence photons is highly dependent on the amplitude of the luminescence pulse I_0.

Figure 24 gives the damping curves of the absorption signal calculated by formula (37) for two photoionization probabilities: $W_1 = \sigma_i I_1 = 3 \cdot 10^5$ sec^{-1} and $W_2 = \sigma_i I_2 = 10^5$ sec^{-1}. It is clear that for a ratio $I_1/I_2 = 3$ the difference in the absorption damping duration is significant. Thus, in order to answer the question of whether or not photoionization of the defects will result in the appearance of the observed absorption pulses it is sufficient to experimentally test the dependence of the duration of damping of the observed absorption on the luminescence intensity or the pump intensity. Experiment reveals that a change in pump intensity by an order of magnitude in the 2-20 MW/cm^2 range has no influence on the duration of the observed absorption, i.e., the photoionization processes of the defects by luminescence are not responsible for the observed signals.

An additional argument supporting this position is the fact that relation (37) is not a simple exponent and will not be straightened out in semilogarithmic coordinates unlike the experimentally-observed absorptions and luminescence (see Figure 23).

3. Tunnel recombination. The processes of tunnel transitions between the trap centers in alkali-halide crystals were discussed in Chapter 2 which demonstrated that these are characterized by a strong (exponential) dependence on the distance between the centers (see formula (6)) or in the final analysis on the defect concentration. The dependence of the duration of damping of the observed absorption on the concentration of activator centers has been investigated experimentally in the $C = 10^{17} - 10^{19}$ cm^{-3} concentration range and in this range no significant dependence of the damping time of absorption on the concentration was observed. This means that the tunnel processes cannot be responsible for the observed absorption signals.

4. Absorption from the relaxed excited states of the activator. The set of experimental facts allows us to interpret the detected absorption as absorption from the relaxed excited states of the In^+ ions. The temperature relations provide conclusive evidence as well as the independence of the observed phenomena of the excitation intensity and activator concentration discussed above. The optical density of absorption from the excited state corresponds to a quantity that would be expected based on the cross-sections measured by the luminescence method. Thus, the mean optical density of absorption for the test crystals was $D \sim 0.2$ with an excited level population of $N \sim 10^{17}$ cm^{-3} and a layer thickness of approximately 1 mm. This corresponds to cross-sections of the phototransitions from the excited state of $\sigma = \alpha/N \sim 2 \cdot 10^{17}$ cm^2 which is in good agreement with the cross-section values given in Table 1.

10. Band States of Activated Alkali-Halide Crystals

The absorption spectrum from the excited state was investigated using the same experimental assembly used in investigating the absorption spectrum of defects formed after the excitation pulse. The measurements were carried out at points corresponding to the spectral lines of a mercury arc lamp. The range of spectral measurements was limited to the shortwave region of the spectrum due to the strong absorption of the probe light (A- and particularly C-bands). Here we may note that luminescence measurements of the quantum efficiency provide good additional information to absorption measurements, since they make it possible to carry out spectral investigations of photoionization in the activator absorption region.

The spectral relations were recorded for crystals with an approximately identical activator concentration $C \sim 10^{18}$ cm^{-3}. The total spectrum for room temperature and a low temperature (T = 100 K) is shown in Figure 25. A strong temperature dependence of the crystals is clear which evidently is due to electron-phonon interaction. From our viewpoint the more important result is the existence at T = 100 K of clearly-expressed maxima in the absorption spectra from the relaxed excited states. This is particularly well manifest for the KBr-In crystal.

Thus, as in the case of luminescence research the absorption spectra from the relaxed excited state reveal the existence of photoionization resonances in the 3.5–4 eV range. The structure of the impurity centers in the photoionozation spectrum may be attributed to many factors discussed previously in § 8. These include the features of the state densities, quasilocal activator levels and the possible influence of upper-lying conduction bands. In order to interpret the observed resonances we require a comprehensive analysis of experimental results incorporating data on the real band structure of alkali-halide crystals.

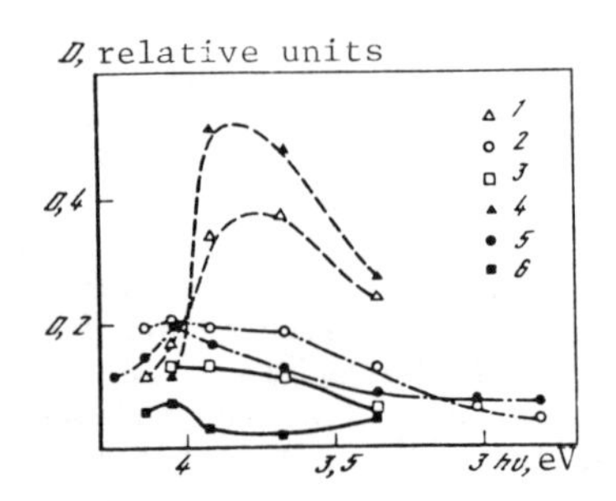

Figure 25. Absorption spectra from the excited state of In^{+} ions in alkali-halide crystals at T = 293 K (1-3) and T = 100 K (4-6).

1, 4 - KBr; 2, 5 - KI; 3, 6 - KCl.

We will focus our evaluation on such important parameters as the minimal photoionization energy E_1, the energy gap between the bottom of the conduction band and the absorption peak (i.e., the band electron energy E_k) and the energy gap between the ground level of the activator and the absorption peak E_2. For this we require the value of Stokes shift E_{CT} in the $^3P_1$-state; this may be estimated based on the single-oscillator harmonic approximation [1] (see also § 1).

We know that this approximation correctly describes the majority of experimental facts in the spectroscopy of mercury-like ions in alkali-halide crystals [1], particularly the Stokes losses. As indicated by the ratio given above the Stokes losses in the harmonic approximation are dependent on the shift between the minima of the potential curves of the ground state and the excited state and on their slope characterized by the force constants α_1 and α_2 [1]:

$$E \text{ and } E' = \frac{1}{2}(\alpha_1 + \alpha_2)x_0^2, \tag{38}$$

where $E^{\text{ж}}$ and E^{I} are the absorption and emission quantum energies.

Each force constant determines the Stokes losses in its "own" state, and therefore it is easily demonstrated that the Stokes losses in the $^3P_1$-state of interest to us may be expressed in the following manner:

$$E_{\text{ст}} = (E^{\text{ж}} - E^{I})(\alpha_1/\alpha_2 + 1)^{-1}. \quad (39)$$

In accordance with [11] Table 2 gives the necessary parameters for indium phosphors and the values of the Stokes shifts in the $^3P_1$-state determined by formula (39).

If we know the Stokes losses in the $^3P_1$-state we may determine the energy gap E_2 between the ground state and the resonance in the conduction band from the simple relation

$$E_2 = E^{\text{ж}} - E_{\text{ст}} + E_{\text{пик}}, \quad (40)$$

where $E_{\text{пик}}$ is the energy of the absorption peak from the excited state.

Table 2

Spectroscopic Parameters of Indium Phosphors

Crystal	$E^{\text{ж}}(^1P_1)$, eV	$\overline{E^{\text{ж}}(^3P_1)}$, eV	$E^{I}(^3P_1)$, eV	α_1/α_2	$E_{\text{ст}}(^3P_1)$, eV
KCl-In	5.46	4.38	2.89	1.6	0.56
KBr-In	5.09	4.23	2.86	1.8	0.49
KI-In	4.73	3.96	2.17	2.0	0.59

Table 3

Energies of the Quasilocal Levels and Band Electrons in Indium Phosphors

Crystal	E_k, eV	E_1, eV	E_2, eV	E_S, eV	E_D, eV
KCl-In	2.46	1.64	7.92	7.3	7.5
KBr-In	2.45	1.35	7.54	7.1	7.4
KI-In	2.74	1.36	7.47	6.5	7.0

It is somewhat more difficult to determine the threshold photoionization energy and the electron energy in the conduction band E_k, since the exact position of the bottom of the conduction band with respect to the $^3P_1$-state is unknown. Available data on thermal delocalization reveal that the energy gap between the bottom of the conduction band and the $^1P_1$-state of mercury-like ions in alkali-halide crystals does not exceed 0.1-0.15 eV [26]. Since thermal ionization of the impurity centers excited to the $^1P_1$-state originates from the relaxed state [26], then by accounting for the Stokes shift in the $^1P_1$-state we may assume with good accuracy (± 0.15 eV) that the energy gap between the ground states of the mercury-like activator and the bottom of the conduction band is equal to the quantum energy of absorption in the C-band $E^{æ}(^1P_1)$. From here

$$E_k = E_2 - E^{æ}(^1P_1), \tag{41}$$

$$E_1 = E_{\text{пик}} - E_k. \tag{42}$$

The quantities E_1, E_k and E_2 calculated by formulae (40)-(42) for the In^+ ions in alkali-halide crystals are shown in Table 3. This same table shows the energy gaps between the ground and highly-excited S- and D-states of the In^+ ions in the alkali-halide crystals estimated by the "linear approximation" [1]. These levels will be discussed in greater detail below.

Now since we have determined certain photoionization parameters we will continue our discussion of their spectral investigations.

The band electron energy $E_k \sim 2.5$ eV which is significant compared to the thermal energy $kT = 0.026$ eV ($T = 300$ K) stands out in Table 3. The quasiwave vector $k_2 = \hbar^{-1} (2m^*E)^{1/2} \sim 10^8$ cm^{-1} corresponds to such an energy with characteristic values of the effective electron mass in the alkali-halide crystals $m^* \sim 0.5\, m$ [65]. The characteristic dimensions of electron localization $æ$ in the A-state are of the order of the lattice constant, i.e., it does not exceed a few angstroms, while the corresponding quasiwave vector of the localized electron does not exceed $k_1 \leqslant æ^{-1} \leqslant$ $\leqslant 3 \cdot 10^7$ cm^{-1}. Selection rule (29) in the photoionization of the localized centers in the crystals forbids optical transitions to an unperturbed conduction band (22) with such a great difference between the quasiwave vectors in the band and localized state ($k_2/k_1 \sim 3$) and the probability of such transitions according to (25) diminishes as $W_{12} \sim k_2^{-6}$, i.e., when $k_2 \sim 10^8$ cm^{-1} it is negligible. Hence it may be stated that singularities exist in our case in the conduction band near $E_k = 2.5$-2.7 eV. In activated alkali-halide crystals the singularities near these values of E_k include possible critical points (or van Hove singularities) [109, 117], d-bands lying above the s-bands [65] and quasilocal states of the activator.

At the critical points the combined (or band-to-band) state density $\rho_{\text{комб}}$ have analytic singularities; here

$$|V_c(p) - V_d(p)| = 0, \tag{43}$$

where $V_c(\mathbf{p})$ and $V_v(\mathbf{p})$ are the electron velocities with quasimomentum $\mathbf{p}$ in the conduction band and the valence band. At the first order critical points, i.e., when

$$V_c(p) = 0; \; V_v(p) = 0, \tag{44}$$

and not only the combined state density but also the separate state densities in the valence band and the conduction band have singularities [109]. Of particular interest are possible critical points related to the overlap of several bands, since for these points the selection rules on quasiwave vector (29) may be satisfied. In other words transitions from the activator relaxed excited state to the upper-lying *d*-conduction band of the alkali-halide crystals are entirely possible. We should note that the critical points for which selection rule (29) is not satisfied will not have a significant influence on photoionization from the localized states, since the growth of the state density near the van Hove singularities obeying the root relation $\rho \sim (E-E_0)^{1/2}$ [109] does not compensate the significantly more rapid decay in the transition probability with growth of energy of the band electron [107, 112].

Evidently the transitions to the *d*-conduction band are initiated at energies exceeding the bandgap by 1.5-2 eV in KCl and 1.3-1.5 eV in KBr [65]. Even if we assume that the transitions to the *d*-band begin from the "ceiling" of the valence band, in this case the energy position of the observed resonances (see E_k in Table 3) is approximately 1-2 eV greater than the *d*-band with respect to the bottom of the *s*-band. Moreover, the sharp shortwave decay in absorption from the excited state in KBr (see Figure 25) is quite uncharacteristic of absorption from the local state to the unperturbed band [108, 112]. We still must recognize that the detected photoionization resonances are related to the quasilocal states of the activator. The following set of results obtained from experiments and their discussion supports this contention.

1. The cross-section of photoionization by quanta with energy E = 4 eV is more than an order of magnitude greater than the cross-section of ionization by quanta with E = 4.6 eV [44], i.e., the transition from the relaxed excited state to the band state in the 4 eV range is a resonance transition.

2. In the spectral range where the absorption maxima from the relaxed excited state are observed, the photoionization cross-sections are close to the cross-sections of the allowed transitions [29, 32].

3. The selection rule for the quasiwave vector (29) is satisfied for the quasilocal levels of the activator regardless of their position with respect to the floor of the conduction band.

4. An analysis of the singularities of the conduction band of the crystal-matrix reveals that the position of the singularities with respect to the floor of the *s*-conduction band of the alkali-halide crystals does not coincide with the position of the photoionization resonances.

The transitions to the upper S and D-levels of mercury-like ions have the highest probability of transitions from the $^3P_1$-level. Hence it is of interest to determine the position of these levels with respect to the ground level for comparison to the value of E_2 given in Table 3. An exact determination of the transition energy to the high-energy states of an impurity ion in the lattice is a difficult theoretical task and has not yet been performed. An approximate estimate of the energy of such transitions was produced in [1, 12] using the so-called compression coefficients

$$\gamma = E_c/E_p, \tag{45}$$

where E_c is the transition energy in a free ion; E_p is the energy of the same transition in a crystalline lattice. The positions of the A- and C-bands in the activated crystals were used to estimate the $\gamma(E_p)$ relation that is approximately described by the function [1]

$$\gamma = 1 + aE_p, \tag{46}$$

i.e., the so-called linear approximation was used according to which the energy levels in the lattice field converge more rapidly than in proportion to their energy. Using the data [11] on the compression coefficients for the $^3P_1$ and $^1P_1$-states we may easily determine the analogous coefficients for the S- and D-levels of the In^+ ion in the alkali-halide crystals (the problem is reduced to drafting line (46) based on two given points) and to determine the energy gaps $E_S$and E_D (see Table 3) between the ground state and the s- and D-states of the impurity ion. It follows from Table 3 that the linear approximation estimates are somewhat reduced compared to the observed resonances in the crystals. The purpose of such an estimate is to demonstrate that the existence of the quasilocal states in this range of values of E_2 has a real basis. Indeed, the group of S-, D- and P-levels that are not manifest as local levels in the bandgap and, consequently, will overlap with the conduction band of the primary material is located above the $^1P_1$-level in accordance with the energy diagrams of the free mercury-like ions demonstrated in [12]. The photoionization resonances discovered in the present study represent the first experimental confirmation of the existence of the quasilocal states of mercury-like ions in alkali-halide crystals.

In our view the present research is complete in some sense: The effective ionization of mercury-like ions in alkali-halide crystals under intense UV was detected; the ionization mechanism was established and the color centers were identified; the photoionization cross-sections of the impurity centers from the excited states were measured and it was demonstrated that the final (band) states of the photoionization transitions are most probably related to the quasilocal states of the activator. Moreover it is of interest to continue the investigation of the interaction between intense UV emission and the activated wideband dielectrics. Several possible areas may be identified here.

1. Further investigations (by the EPR technique) of the structure of defects formed in crystals under optical excitation, and an investigation of the optical characteristics of defects and a search for laser-active centers.

2. Expanding the spectral and temporal range of research on the spectroscopy of the excited states of impurity centers as well as an experimental and theoretical investigation of the band structure of activated dielectrics and self-ionization resonances in crystals.

3. Expanding intense excitation sources (lasers) to the vacuum UV range and formulating new experiments employing such lasers.

4. Applied research on the application of the effective generation of color centers in crystals by UV excitation in optical data recording devices.

CONCLUSION

The following are the primary results of the present study.

1. Effective ionization of mercury-like ions (In^+, Tl^+, Sn^{2+}) in alkali-halide crystals under intense ($I \sim 10^6$ W/cm^2) optical excitation in the A-absorption band of the impurity ions was discovered. The ionization probability is $10^6 - 10^7$ sec^{-1}.

2. It was established experimentally that the mechanism responsible for the ionization of the mercury-like ions in alkali-halide crystals under optical excitation in the A-band is stepwise photoionization of the activator. Photoionization of the impurity ions is accompanied by the formation of color centers in the crystals as well as recombination luminescence and photoconductivity.

3. Joint EPR and optical techniques were used to demonstrate that under selective optical excitation of alkali-halide crystals with mercury-like ions (to the activator absorption bands) impurity centers are preferentially created in the crystals.

4. A technique is proposed for measuring the photoionization cross-sections from the excited state of the impurity centers in crystals by the change in relative quantum efficiency of intracenter luminescence. The photoionization cross-sections from the lower excited state of the In^+, Tl^+ and Sn^{2+} ions were measured in alkali-halide crystals ($1\text{-}4 \cdot 10^{-17}$ cm^2); these were close to the cross-sections of the forbidden electron dipole transitions.

5. Resonances to the conduction band were detected in the absorption spectrum of the excited state of mercury-like ions in alkali-halide crystals. An analysis of the experimental data and the features of the band structure of the alkali-halide crystals revealed that the observed resonances may be attributed to the existence of quasilocal states of the activator.

The author wishes to express his deep gratitude to Academician A. M. Prokhorov and Doctor of Physics and Mathematics T. M. Murina for their scientific leadership and constant assistance in this study. The author is genuinely grateful to V. I. Zhekov, P. G. Baranov and L. E. Nagli for their scientific collaboration, assistance in experiments and helpful discussions. The author is also grateful to M. V. Fedorov and A. E. Kazakov for useful discussions and to the entire staff of the Laboratory of Oscillation Research of the Institute of General Physics of the Academy of Sciences of the USSR who provided assistance in this study.

BIBLIOGRAPHY

1. Lushchik, N. E., Lushchik, Ch. B. A Model of Luminescence Centers in Alkali-Halide Crystals. Tr. In-ta fiziki i astronomii AN ESSR, 1957, No. 6, pp. 5-62.

2. Lushchik, Ch. B., Lushchik, N. E., Zazubovich, S. G. et al. Rtutepodobnye tsentry lyuminestsentsii v ionnykh kristallakh [Mercury-Like Luminescence Centers in Ionic Crystals]. Fizika shchelochno-galoidnykh kristallov [The Physics of Alkali-Halide Crystals]. Edited by K. K. Shvarts. Riga: Latv. un-t, 1962, pp. 102-115.

3. Lushchik, N. E., Meriloo, I. A. Phosphors With Mercury-Like Activators and the Problem of Predicting the Spectra of New Luminophores. Izv. AN SSSR. Ser. Fiz., 1966, Vol. 30, No. 9, pp. 1517-1520.

4. Hilsch, R. Die Absorptionsspektra einiger Alkali-Halogenidphosphore mit Tl- und Pb-Zusatz. Ztschr. Phys., 1927, Vol. 44, pp. 860-870.

5. Hilsch, R., Pohl, R. W. Uber die ersten ultravioletten Eigenfrequenzen einiger einfacher Kristalle. Ztschr. Phys., 1928, Vol. 48, pp. 384-396.

6. Vyazemskiy, V. I., Lomonosov, I. I., Pisarevskiy, A. N. et al. Stsintillyatsionnyy metod v radiometrii [The Scintillation Method in Radiometry]. Moscow: Gosatomizdat, 1961, 430 pages.

7. Svenk, R. K. The Characteristics of Scintillators. UFN, 1956, Vol. 58, pp. 519-553.

8. Gellerman, W., Koch, K. P., Luty, F. Recent Progress in Color Center Lasers. Laser Focus, 1982, Vol. 18, pp. 71-75.

9. Vlasov, G. I., Kalnyn'sh, R. A., Nagli, L. E. et al. Certain Physical Phenomena in Activated Alkali-Halide Crystals and the Possibilities for Optical Data Processing. Avtometriya, 1980, No. 1, pp. 66-96.

10. Zazubovich, S. G., Lushchik, N. E. Elektronnye vozbuzhleniya rtutepodobnykh tsentrov v shchelochnogaloidnykh kristallakh [Electron

Excitations of Mercury-Like Ions in Alkali-Halide Crystals]. Fizika primesnykh tsentrov v kristallakh [The Physics of Impurity Centers in Crystals]. Edited by G. S. Zavt. Tallin: Izd-vo AN ESSR, 1972, pp. 483-504.

11. Lushchik, N. E. Indium-Activated Alkali-Halide Phosphors. Tr. In-ta fiziki i astronomii AN ESSR, 1957, No. 6, pp. 149-168.

12. Lushchik, N. E., Lushchik, Ch. B., Liyd'ya, G. G. et al. Localized Electron Excitations of Ionic Crystals Activated by Mercury-Like Ions. Tr. In-ta fiziki i astronomii AN ESSR, 1964, No. 28, pp. 3-19.

13. Fukuda, A. Alkali-Halide Phosphors Containing Impurity Ions With $(s)^2$ Configuration. Sci. Light, 1964, Vol. 13, pp. 64-114.

14. Fowler, W. B. Electronic States and Optical Transitions of Color Centers. Physics of Color Centers. Ed. W. B. Fowler. N. Y.: Acad. Press, 1968, pp. 133-150.

15. Trinkler, M. F. The Complex Structure of the Lower Excited State of Thallium-Activated Alkali-Halide Crystals. Radiatsionnaya fizika. Riga: Zinatie, 1970, No. 6, pp. 6-47.

16. Tale, A. K. The Intracenter Luminescence of Indium-Activated Alkali-Halide Crystals. Radiatsionnaya fizika. Riga: Zinatie, 1970, No. 6, pp. 49-92.

17. Kristofel', N. N. Teoriya primesnykh tsentrov malykh radiusov v ionnykh kristallakh [The Theory of Small Radius Impurity Centers in Ionic Crystals]. Moscow: Nauka, 1974, 336 pages.

18. Aseev, G. I., Kats, M. L., Nikol'skiy, V. K. et al. Multiphoton Excitation of Luminescence and Photoconductivity of KCl-Eu Monocrystals by Ruby and Neodymium Lasers. Izv. AN SSSR. Ser. fiz., 1969, Vol. 33, No. 5, pp. 857-862.

19. Aseev, G. I., Kats, M. L., Nikol'skiy, V. K. et al. Accounting For Saturation in Multiphoton Ionization Processes of Activated Alkali-Halide Crystals. FTT, 1974, Vol. 16, Issue 1, pp. 293-295.

20. Aseev, G. I., Kats, M. L., Nikol'skiy, V. K. et al. The Mechanisms of the Reduction in Ionization Potential of Impurity Centers in Alkali-Halide Crystals in a Laser Field. Optika i spektroskopiya, 1975, Vol. 38, pp. 959-965.

21. Baranov, P. G., Danilov, V. P., Zhekov, V. I. et al. Opticheskoe pogloshchenie kristallov $KI-Tl^+$, navedennoe intensivnym UF lazernym izlucheniem [Optical Absorption of $KI-Tl^+$ Crystals Induced by Intense UV Laser Emission]. 27-e Vsesoyuz. soveshch. po lyuminestsentsii [The 27th All-Union Conference on Luminescence]. Ezernieki, Latvian SSR, 12-16 May 1980. Topic Papers. Riga: The Institute of Physics of the Academy of Sciences of the Latvian SSR, 1980, p. 101.

22. Baranov, P. G., Danilov, V. P., Zhekov, V. I. et al. Induced Optical Absorption of KI-Tl Crystals Under Intense Laser Irradiation. Krat. soobshch. po fizike FIAN, 1980, No. 5, pp. 33-38.

23. Baranov, P. G., Danilov, V. P., Zhekov, V. I. et al. Optical Absorption of KI-Tl Crystals Induced by Intense UV Laser Emission. FTT, Vol. 22, Issue 9, pp. 2790-2796.

24. Baranov, P. G., Danilov, V. P., Zhekov, V. I. et al. Formation of Color Centers in KCl-In and NaCl-In Crystals Under Intense UV Laser Irradiation. FTT, 1981, Vol. 23, Issue 6, pp. 1829-1831.

25. Tiysler, E. S. Optical and Electric Phenomena in the Delocalization of Impurity Excitations in Ionic Crystals. Izv. AN SSSR. Ser. fiz., 1966, Vol. 30, No. 9, pp. 1545-1548.

26. Tiysler, E. S. Optical and Electrical Phenomena in the Ionization of Luminescence Centers in Alkali-Halide Crystals Activated by Ga^+, In^+ and Tl^+ Ions. Dissertation For Candidate of Mathematics and Physics. Tartu, 1968, 19 pages.

27. Lushchik, Ch. B., Kyazmbre, Kh. F., Lukantsever, Yu. L. et al. Ionization of Luminescence Centers and Quasilocal Electron Excitations in Ionic Crystals. Izv. AN SSSR. Ser. fiz., 1969, Vol. 33, No. 5, pp. 863-869.

28. Vishnevskiy, V. N., Pidzyraylo, N. S. Recombination Luminescence of Nai-Tl Monocrystals Excited in the A-Activator Absorption Band. Izv. vuzov. Fizika., 1972, No. 7, pp. 158-160.

29. Danilov, V. P., Zhekov, V. I., Murina, T. M. et al. Laser Ionization of Mercury-Like Ions in Alkali-Halide Crystals. Izv. AN Latv. SSR. Ser. fiz. i tekhn. nauk., 1982, No. 3, pp. 44-46.

30. Lusis, D. Yu. Luminescence and Energy Storate in KCl-In Under Excitation in the Fundamental Absorption Range. Izv. AN Latv. SSR. Ser. fiz. i tekhn. nauk., 1974, No. 6, pp. 35-40.

31. Kristofel', N. N., Zavt, G. S. Optical Electron Transitions Between Local and Band States in Impurity Centers. Optika i spektroskhopiya, 1968, Vol. 25, pp. 705-712.

32. Danilov, V. P., Zhekov, V. I., Murina, T. M. et al. The Cross-Sections of Photoionization From the Excited State of Certain Mercury-Like Ions in Alkali-Halide Crystals. Kvantovaya elektron., 1982, Vol. 9, No. 7, pp. 1474-1478.

33. Fano, U. Effects of Configuration Interaction on Intensities and Phase Shifts. Phys. Rev., 1961, Vol. 124, pp. 1866-1878.

34. Fano, U., Kuper, Dzh. Spektral'nye raspredeleniya sil ostsillyatorov v atomakh [The Spectral Distributions of Oscillator Forces in Atoms]. Moscow: Nauka, 1972, 200 pages.

35. Lunch, D. W., Robinson, D. A. Study of the F'-Center in Several Alkali-Halides. Phys. Rev., 1968, Vol. 174, pp. 1050-1059.

36. Delbecq, C. J., Ghosh, A. K., Yuster, P. H. Trapping Annihilation of Electrons and Positive Holes in KCl:TlCl. Phys. Rev., 1966, Vol. 151, pp. 599-609.

37. Baranovski, J. M., Langer, J. M. Observation of Discrete Excited States Degenerate With Conduction Band in CdTe With Ti and CdSe With Co Impurities. 11-th Intern. Conf. Physics of Semiconductors. Abstr. W-wa, 1972, pp. 137-138.

38. Abagyan, S. A., Ivanov, G. A., Kuznetsov, Yu. N. et al. The Excited Levels of a Local Center in Resonance With Band Conductivity: Cr-GaP and GaAs. Fizika i tekhnika poluprovodnikov, 1973, Vol. 7, Issue 8, pp. 1474-1478.

39. Ippolitova, G. K., Omel'yanovskiy, E. M., Pervova, L. D. Intracenter Optical Transitions in GaAs: Cr With Continuum Resonance. Fizika i tekhnika poluprovodnikov, 1975, Vol. 9, Issue 7, pp. 1308-1313.

40. Danilov, V. P., Zhekov, V. I., Murina, T. M. et al. Vysokoenergeticheskie kvazilokal'nye sostoyaniya v aktivirovannykh shchelochno-galoidnykh kristallakh [High-Energy Quasilocal States in Activated Alkali-Halide Crystals]. 6-ya Vsesoyuz. konf. po fizike VUF-izlucheniya i vzaimodeystviyu izlucheniya s veshchestvom [The 6th All-Union Conference on the Physics of Vacuum UV Emission and the Interaction of Emission and Matter]. Moscow, 22-24 June 1982 g.: Tez. dokl. Moscow: Izd-vo MGU, 1982, p. 117.

41. Danilov, V. P., Zhekov, V. I., Murina, T. M. et al. Quasilocal States in Indium-Activated Alkali-Halide Crystals. Pis'ma v ZhETF, 1982, Vol. 36, Issue 6, pp. 187-189.

42. Danilov, V. P., Zhekov, V. I., Murina, T. M. et al. Dvukhfotonnaya stupenchataya spektroskopiya kvazilokal'nykh sostoyaniy rtutepodobnykh ionov v shchelochno-galoidnykh kristallakh [Two-Photon Stepwise Spectroscopy of Quasilocal States of Mercury-Like Ions in Alkali-Halide Crystals]. 11-ya Vsesoyuz. konf. po kogerentnoy i nelinejnoy optike [The 11th All-Union Conference on Coherent and Nonlinear Optics]. Erevan, 22-25 November 1982. Topic Papers. Erevan: Erevanskiy gos. un-t, 1982, Ch. 1, pp. 255-256.

43. Danukov, V. P., Zhekov, V. I., Murina, T. M. et al. Vysokoenergeticheskie sostoyaniya v aktivirobannykh shchelochno-galoidnykh kristallakh [The High-Energy States in Activated Alkali-Halide Crystals]. The Physics of Vacuum UV Emission. Tr. 6-y Vsesoyuz. konf. po fizike VUF izlucheniy [Papers of the 6th All-Union Conférence on the Physics of Vacuum UV Emission]. Moscow: Izd-vo MGU, 1984, pp. 103-107.

44. Danilov, V. P., Zhekov, V. I., Murina, T. M. et al. The Photoionization of Mercury-Like Ions in Alkali-Halide Crystals Under Excitation

in the A- and C-Bands. Krat. coobsh. po fizike FIAN, 1982, No. 7, pp. 25-28.

45. Letokhov, V. S. The Problems of Laser Spectroscopy. UFN, 1976, Vol. 118, pp. 199-249.

46. Lazernaya spektroskopiya atomov i molekul [The Laser Spectroscopy of Atoms and Molecules]. Edited by G. Val'ter. Moscow: Mir, 1979, 432 pages.

47. Mel'nikov, N. I., Baranov, P. G., Zhitnikov, R. A. et al. Anisotropic Silver Centers in a KCl Crystal. FTT, 1971, Vol. 13, Issue 8, pp. 2276-2282.

48. Gellerman, W., Luty, F., Pollock, C. R. Optical Properties and Stable, Broadly Tunable CW Laser Operation of New F_a-Type Centers in Tl^+-Doped Alkali-Halides. Opt. Communs., 1981, Vol. 39, pp. 391-395.

49. Seitz, F. Interpretation of the Properties of Alkali-Halide Thallium Phosphors. J. Chem. Phys., 1938, Vol. 6, pp. 150-162.

50. Lushchik, N. E., Lushchik, Ch. B. Electron-Vibrational Processes in the Luminescence Centers of Ionic Crystals Involving Several Excited States. Tr. In-ta fiziki i astronomii AN ESSR, 1961, No. 15, pp. 30-55.

51. Edgerton, R., Teegarden, K. Emission Spectra of KI-Tl at 12 K. Phys. Rev., 1964, Vol. 136, pp. 1091-1092.

52. Sil'd, O. I. Teoriya tsentra lyuminestsentsii v kristalle [The Theory of Luminescence Centers in Crystals]. Tartu: Izd-vo Tart. un-ta, 1968, 138 pages.

53. Rebane, K. K. Elementarnaya teoriya kolebatel'noy struktury spektra primesnykh tsentrov kristallov [Elementary Theory of the Vibrational Structure of the Spectrum of Impurity Centers in Crystals]. Moscow: Nauka, 1968, 232 pages.

54. Trinkler, M. F., Zolovkina, I. S. A-Luminescence of Alkali-Halide Crystals Activated by Single-Valence Mercury-Like Ions. Izv. AN SSSR. Ser. fiz., 1976, Vol. 40, No. 9, pp. 1939-1943.

55. Fukuda, A. Effekt Yana-Tellera i tsentry tipa Tl^+ v shchelochno-galoidnykh kristallakh [The Jahn-Teller Effect and Tl^+ Centers in Alkali-Halide Crystals]. Fizika primesnykh tsentrov v kristallakh [The Physics of Impurity Centers in Crystals]. Edited by G. S. Zavt. Tallin: Izd-vo AN ESSR, 1972, pp. 505-527.

56. Le Si Dang, Romestain, R., Merle d'Aubigne, Y. et al. Jahn-Teller Effects in an Orbital Triplet Coupled to Both E_g and T_{2g} Modes of Vibration: Experimental Evidence For the Coexistence of Tetragonal Minima. Phys. Rev. Lett., 1977, Vol. 38, pp. 1539-1543.

57. Le Si Dang, Merle d'Aubigne, Y., Romestain, R. et al. Magnetic Resonance in Relaxed States A_x and A_t of Ga^+ in Alkali-Halides. Solid State Communs., 1978, Vol. 26, pp. 413-416.

58. Trinkler, M. F. The Static Jahn-Teller Effect For Mercury-Like Luminescence Centers in Alkali-Halide Crystals. Izv. AN SSSR. Ser. fiz., 1981, Vol. 45, No. 2, pp. 332-336.

59. Zazubovich, S. G. Investigations of the Structure of the Excited States of Mercury-Like Centers in Cubic Crystals by the Polarization Luminescence Technique. Izv. AN SSSR. Ser. fiz., 1982, Vol. 46, No. 2, pp. 273-279.

60. Lushchik, Ch. B. Investigation of Trap Centers in Alkali-Halide Crystals. Tr. In-ta fiziki i astronomii AN ESSR, 1955, No. 3, 230 pages.

61. Mott, N., Gerni, R. Elektronnye protsessy v ionnykh kristallakh [Electron Processes in Ionic Crystals]. Moscow: Izd-vo inostr. lit., 1950, 304 pages.

62. Kats, M. L. Lyuminestsentsiya i elektronno-dyrochnye protsessy v fotokhimicheski okrashennykh kristallakh chehelochno-galoidnykh coedineniy [The Luminescence and Electron Hole Processes in the Photochemically Colored Crystals of Alkali-Halide Compounds]. Saratov: Izd-vo Sarat. un-ta, 1960, 271 pages.

63. Gurvich, A. M. Vvedenie v fizicheskuyu khimiyu kristallofosforov [An Introduction to the Physical Chemistry of Crystal Phosphors]. Moscow: Vyssh. shk., 1971, 356 pages.

64. Lushchik, Ch. B., Vitol, I. K., Elango, M. A. The Decay of Electron Excitations Into Radiation Defects in Ionic Crystals. UFN, 1977, Vol. 122, pp. 223-251.

65. Alker, E. D., Lusis, D. Yu., Chernov, S. A. Elektronnye vozbuzhdeniya i radiolyuminestsentsiya shchelochno-galoidnykh kristallov [Electron Excitations and the Radio Luminescence of Alkali-Halide Crystals]. Riga: Zinatie, 1979, 251 pages.

66. Stounkhem, A. M. Teoriya defektov v tverdykh telakh [Defect Theory in Solids]. Moscow: Mir, 1978, Vol. 1, 569 pages; Vol. 2, 357 pages.

67. Vale, G. K. Electron and Hole Trapping by Luminescence Centers in Alkali-Halide Crystals Activated by Mercury-Like Ions. Tr. In-ta fiziki i astronomii AN ESSR, 1963, No. 23, pp. 137-154.

68. Vale, G. K., Zolotarev, G. K., Kuketaev, T. A. et al. Activator Traps For Electrons and Holes in Ionic Crystals. Izv. AN SSSR. Ser. fiz., 1966, Vol. 30, No. 4, pp. 695-697.

69. Kyaembre, Kh. F., Okk, M. F., Yaek, I. V. Optical and Thermooptic Electrons and Photostimulated Luminescence in KCl-Tl. Izv. AN SSSR. Ser. fiz., 1966, Vol. 30, No. 9, pp. 1451-1453.

70. Liyd'ya, G. G., Yaek, I. V. The Generation of F-Centers in Alkali-Halide Phosphors by UV Radiation. Tr. In-ta fiziki i astronomii AN ESSR, 1961, No. 14, pp. 212-235.

71. Hersh, H. N. Color Centers in KI containing Impurity Electron Acceptors KI-Tl. J. Chem. Phys., 1959, Vol. 30, pp. 790-800.

72. Lushchik, Ch. B., Tiysler, E. S. A Spectrophotometric Investigation of Delocalization of Excitations in Ionic Crystals. Tr. In-ta fiziki i astronomii AN ESSR, 1960, No. 12, pp. 125-148.

73. Jack, I., Kink, M. Tl^{o}, Ga^{o} and In^{o} Centers in KCl Crystals. Phys. Status Solidi (b), 1973, Vol. 56, pp. 375-381.

74. Zolotarev, G. K., Lushchik, Ch. B., Soovik, T. A. et al. Self-Localization of Holes and Optical Phenomena in Ionic Crystals. Izv. AN SSSR. Ser. fiz., 1965, Vol. 29, No. 1, pp. 36-39.

75. Chernyak, V. G., Plyavin', I. K. Maloinertsionnaya peredacha energin v shchelochno-galoidnykh kristallakh, aktivirovannykh rtutepodobnymi ionami [Low-Inertia Energy Transmission in Alkali-Halide Crystals Activated by Mercury-Like Ions]. Radiatsionnaya fizika [Radiation Physics]. Riga: Zinatie, 1970, No. 6, pp. 93-162.

76. Pung, L. A., Khaldre, Yu. Yu. Investigation of the Electron and Hole Processes in Ionic Crystals Based on Nonisothermic EPR Relaxation and Recombination Luminescence. Tr. In-ta fiziki i astronomii AN ESSR, 1970, No. 38, pp. 50-84.

77. Dreybrodt, W., Silber, D. Electron Spin Resonance of Tl^{2+} Centers in KCL Crystals. Phys. status solidi, 1967, Vol. 20, pp. 337-346.

78. Osminin, V., Zazubovich, S. Processes Due to Optical Excitation of Impurities With One s-Electron (Af^{o}, Cu^{o}, Tl^{2+}, In^{2+}, Ga^{2+}) in KCl Crystals. Phys. status solidi, 1975, Vol. 71, pp. 435-447.

79. Baranov, P. G., Khramtsov, V. A. The EPR of Indium and Gallium Centers in Alkali-Halide Crystals. FTT, 1979, Vol. 21, Issue 5, pp. 1455-1460.

80. Baranov, P. G., Khramtsov, V. A. The EPR of Ga^{+} and Tl^{2+} Ions in KCl and NaCl Crystals. FTT., 1978, Vol. 20, Issue 6, pp. 1870-1875.

81. Baranov, P. G. Microwave-Optical Spectroscopy of Defects in Ionic Crystals: Dissertation For Doctor of Physics and Mathematics. I. A. Ioffe, 1981, 396 pages.

82. Lushchik, Ch. B., Volin, I. V. Recombination Luminescence of Alkali-Halide Crystals Activated by Mercury-Like Ions. Tr. In-ta fiziki i astronomii AN ESSR, 1958, No. 7, pp. 311-339.

83. Lukantsever, N. L. The Upper and Lower Ionization of Mercury-Like Ions in KCl-Ga, Ag; KCl-In, Ag and KCl-Tl, Ag Crystals. ZhPS, 1969, Vol. 11, pp. 309-315.

84. Pologrudov, V. V., Karnaukhov, E. N. O mekhanizme ionizatsii aktivatora pri dlinnovolnovom vozbuzhdenii shchelochno-galoidnykh kristallov [The Ionization Mechanism of an Activator Under Longwave Excitation of Alkali-Halide Crystals]. 6-ya Vsesoyuz. konf. po fizike VUF-izlucheniya i vzaimodeystviyu izlucheniya s veshchestvom [The 6th All-Union Conference on the Physics of Vacuum UV Emission and the Interaction of Emission Matter]. Moscow, 22-24 June 1982: Tez. dokl. Moscow: Izd-vo MGU, 1982, p. 125.

85. Bredikhin, V. I., Genkin, V. N., Miller, A. M., Soustov, L. V. Experimental Investigation of the Nature of Photoelectric Phenomena in KDR and DKDR Crystals. ZhETF, 1978, Vol. 75, Issue 5, pp. 1763-1770.

86. Soustov, L. V. Investigation of the Nature of Photoelectric Effects Under Excitation of Transparent Crystals by Laser Pulses, 1982, 24 pages.

87. Dietrich, H. B., Purdy, A. E., Murray, R. B. et al. Kinetics of Self-Trapped Holes in Alkali-Halide Crystals: Experiments in NaI (Tl) and KI (Tl). Phys. Rev. (b). Solid State, 1973, Vol. 8, pp. 5894-5901.

88. Zazubovich, S. G., King, M. F., Yaek, I. V. et al. The Stepwide Emission of Two Photons From Interaction of a Single Electron-Hole Pair With Activator Centers in Alkali-Halide Crystals. Izv. AN SSSR. Ser. fiz., 1973, Vol. 37, No. 4, pp. 732-735.

89. Baranov, P. G., Veshunov, Yu. P., Zhitnikov, R. A. EPR, Optical Absorption and Luminescence of Anisotropic Copper Centers in KCl. Phys. status solidi (b), 1977, Vol. 79, pp. K27-K30.

90. Kulis, P. A. Parametry teplovoy ionizatsii atomarnykh tsentrov aktivatora v shchelochnogaloidnykh kristallakh [Parameters of the Thermal Ionization of Atomic Activator Centers in Alkali-Halide Crystals]. Uchen. zap. Latv. un-ta, 1975, Vol. 246, pp. 111-126.

91. Gorshkov, B. G., Epyfanov, A. S., Manenkov, A. A. et al. An Experimental Investigation of Photoconductivity of Broadband Dielectrics Excited by UV Laser Emission. ZhETF, 1981, Vol. 81, Issue 4, pp. 1423-1434.

92. Smakula, A. Uber Erregung und Entfarbung lichtelektrischleitender Alkalihalogenide. Ztschr. Phys., 1930, Issue 59, pp. 603-614.

93. Dexter, D. L. Absorption of Light by Atoms in Solids. Phys. Rev., 1956, Vol. 101, pp. 48-55.

94. Feofilov, P. P. Kooperativnye opticheskie yavleniya v aktivirovannykh kristallakh [Cooperative Optical Phenomena in Activated Crystals]. Fizika primesnykh tsentrov v kristallakh [The Physics of Impurity Centers in Crystals]. Edited by G. S. Zavta. Tallin: Izd-vo AN ESSR, 1972, pp. 539-563.

95. Leyman, V. I. Tunnel Ionization of Excited In^{+}-Centers in KCl-In. FTT, 1972, Vol. 14, Issue 12, pp. 3650-3654.

96. Delone, N. B., Kraynov, V. P. Atom v sil'nom svetovom pole [The Atom in a Strong Light Field]. Moscow: Atomizdat, 1978, 287 pages.

97. Keldysh, L. V. Ionization in the Field of a Strong Electromagnetic Wave. ZhETF, 1964, Vol. 47, Issue 5, pp. 1945-1957.

98. Voronkova, E. M., Gretsushnikov, B. N.. Distler, G. I., Petrov, I. P. Opticheskie materialy dlya infrakrasnoy tekhniki [Optical Materials For Infrared Technology]. Moscow: Nauka, 1965, 335 pages.

99. Ovsyankin, V. V., Feofilov, P. P. Cooperative Processes in Luminescent Systems. Izv. AN SSSR. Ser. fiz, 1973, Vol. 37, No. 2, pp. 262-272.

100. Basiev, T. T., Voron'ko, Yu. K., Mirov, S. B. et al. The Kinetics of Accumulation and Generation of F_2^+-Centers in LiF (F_2) Crystals. Pis'ma v ZhETF, 1979, Vol. 30, Issue 10, pp. 661-665.

101. Morgenshtern, Z. L. Measuring the Absolute Quantum Efficiency of Photoluminescence of Alkali-Halide Crystals. ZhETF, 1955, Vol. 29, pp. 903-904.

102. Antonov-Romanovskiy, V. V. Vvedenie v kinetiku fotolyuminestsentsii kristallofosforov [An Introduction to the Kinetics of Phosphor Photoluminescence]. Moscow: Nauka, 1966, 324 pages.

103. Ohata, T., Hayashi, T., Koshino, S. Luminescence of $(Tl^{+})_2$ Centers in KI. J. Phys. Soc. Jap., 1978, Vol. 45, pp. 581-585.

104. Billardon, M., Ortega, J. M. On the Existence of a Reabsorption Between Excited States of "ns^2" Ions in Alkali-Halides. Solid State Communs., 1981, Vol. 38, pp. 765-769.

105. Intenberg, L. E. An Investigation of the Damping of the Fast Impurity Luminescence Component Under Excitation in the A- and C-Absorption Bands in a KI-Tl Crystal. Izv. AN Latv. SSR. Ser. fiz. i tekhn. nauk., 1977, No. 5, pp. 33-38.

106. Milis, A. Primesi s glubokimi urovnyami v poluprovodnikakh [Impurities With Deep Levels in Semiconductors]. Moscow: Mir, 1977, 562 pages.

107. Kireev, P. S. Fizika poluprovodnikov [The Physics of Semiconductors]. Moscow: Vysch. shk., 1975, 584 pages.

108. Pankov, Zh. Opticheskie protsessy v poluprovodnikakh [Optical Processes in Semiconductors]. Moscow: Mir, 1973, 456 pages.

109. Bonch-Bruevich, V. L., Kalashnikov, S. G. Fizika poluprovodnikov [The Physics of Semiconductors]. Moscow: Nauka, 1977, 672 pages.

110. Gribkovskiy, V. P. Teoriya pogloshcheniya i ispuskaniya sveta v poluprovodnikakh [The Theory of Light Absorption and Emission in Semiconductors]. Minsk: Nauka i tekhnika, 1975, 464 pages.

111. Page, L. Y., Strozier, Y. A., Hygh, E. H. Calculation of L Band in KCl. Phys. Rev. Lett., 1968, Vol. 21, pp. 348-350.

112. Kohn, W. Shallow Impurity in Silicon and Germanium. Solid State Phys., 1957, Vol. 5, pp. 267-320.

113. Dzhonson, E. Pogloshchenie vblizi kraya fundamental'noy polosy [Absorption Near the Fundamental Band Edge]. Opticheskie svoystva poluprovodnikov [The Optical Properties of Semiconductors]. Edited by R. Uillardson, A. Bira. Moscow: Mir, 1970, pp. 166-277.

114. Lucovsky, G. On the Photoionization of Deep Impurity Centers in Semiconductors. Solid State Communs., 1965, Vol. 3, pp. 299-302.

115. Shmartsev, Yu. V., Reshetnyuk, A. D., Kistova, E. M. Photoionization of Donors With Electron Transition to the Higher Conduction Band. Fizika i tekhnika poluprovodnikov, 1970, Vol. 4, Issue 1, pp. 195-199.

116. Lushchik, Ch. B. Electron Excitations and Electron Processes in Luminscencing Ionic Crystals. Tr. In-ta fiziki i astronomii AN ESSR, 1966, No. 31, pp. 19-83.

117. Fillips, Dzh. Opticheskie spektry tverdykh tel [The Optical Spectra of Solids]. Moscow: Mir, 1968, 176 pages.

118. Fillips, Dzh. The Optical Properties of Semiconductors. Edited by R. Uillardoson and A. Bir. Moscow: Mir, 1970, pp. 278-312.

119. Mollenauer, L. F. Excited-State Absorption Spectrum of F_2^+ Centers and the H_2^+ Molecular-Ion Model. Phys. Rev. Lett., 1979, Vol. 43, No. 20, pp. 1524-1528.

Laser Excitation of Nonequilibrium Carriers in Wideband Dielectrics

B.G. Gorshkov, A.S. Epifanov, A.A. Manenkov
A.A. Panov

Abstract: The fundamental regularities of the photoexcitation of nonequilibrium carriers are determined based on comprehensive experimental investigations of the photoconductivity, nonlinear absorption and drag effect in updoped alkali-halide crystals irradiated by nanosecond UV laser pulses (at 0.27 and 0.35 μm). A theoretical analysis of the photoconductivity phenomena observed in the alkali-halide crystals is given.

The photoconductivity of KDP and DKDP crystals under laser irradiation at 0.27 μm is discovered and investigated. A reduction in the photoconductivity of ruby exposed to multiple laser UV irradiation of constant intensity was observed; this phenomenon is related to space charge formation.

INTRODUCTION

Phenomena from the interaction of high-power laser emission with transparent solids has been the focus of increasing interest. The importance of such phenomena for fundamental solid-state physics is beyond dispute due to their relation to many problems such as multiphoton transitions, nonequilibrium carrier generation and recombination processes and the formation of radiation defects.

From the applied viewpoint this process is related to the problem of laser-induced damage of the optical materials used as active and passive components in laser equipment.

A number of recent theoretical and experimental studies (see, for example, [1-7]) have identified the principal mechanisms responsible for the breakdown of transparent solid dielectrics in an intense electromagnetic wave field. It has been demonstrated that in the majority of realizable cases damage may be attributed to absorbing inclusions, although in the purest samples, the damage may be attributed to electron processes: Impact and multiphoton ionization. The interaction between multiphoton absorption processes and avalanche breakdown mechanisms in solids was discussed in [8-11]. Recent investigations [12-14] experimentally discovered laser-induced damage due to impact ionization (electron avalanche) in alkali-halide crystals.

In this regard a detailed investigation of the electron processes generating laser-induced damage and determining the maximum laser radiation strength of optical materials is critical. Mention should also be made of such issues as the photoexcitation and recombination mechanisms of non-equilibrium carriers as well as estimates of their concentration at various intensities and frequencies of laser emission; determination of the seed electron sources [12, 14, 15] including the role of multiphoton ionization of the fundamental lattice atoms and the role of impurity and defect ionization.

Among the various experimental techniques for investigating electron processes in solids one of the most effective methods is based on photoconductivity. As a direct method in which nonequilibrium carriers are directly detected, this method makes it possible to obtain information on the excitation and recombination processes as well as carrier concentrations and mobilities [16]. Therefore in the present study we employed this method as the principal method for investigating photoexcitation and carrier recombination mechanisms in wideband dielectrics in the UV.

We note that the phenomenon of photoconductivity has been virtually ignored in investigating the mechanisms of optical destruction of transparent dielectrics. We are aware of only two studies [17, 18] where in a very narrow range of laser emission intensities (virtually at the optical breakdown threshold itself) volumetric photoconductivity was observed in ruby at 0.69 μm and a much later study [19] where the dynamics of laser-induced damage of a KDP crystal was discovered at the wavelength of a neodymium laser from a photoresponse investigation.

In order to more reliably interpret the results presented in this study on the photoconductivity and other related conclusions with respect to photoexcitation and the recombination of nonequilibrium carriers we investigated (in alkali-halide crystals in the UV) the photon drag effect that had not been observed previously in dielectric crystals together with the nonlinear absorption of laser emission.

In order to investigate the features of the drag effect we developed an original technique that makes it possible to measure the field strength of the effect and to obtain its dependence on the laser emission intensity. By using the optoacoustic method in our experiments which is a high sensitivity method the nonlinear absorption of energy attributed to the multiphoton transitions could be observed at emission intensity levels several orders of magnitude lower than the threshold for laser-induced damage [12, 20].

We should emphasize that the investigations of the photon drag effect and nonlinear absorption were carried out for the same wavelengths and on the same samples as in the case of photoconductivity investigations which made it possible to correctly correlate the results from these independent investigations and therefore draw substantiated conclusions regarding the photoexcitation and carrier recombination mechanisms.

Another problem related to the interaction of high-power laser emission with solid transparent dielectrics is related to investigations of radiation defects and tinted centers. Determining the formation mechanisms of defects under the influence of high-power laser emission is of significant practical interest both for fundamental solid-state physics and for many practical applications. Specifically, such defects may play a significant role in the degradation processes of laser elements (active and passive) and may influence the kinetics of lasing and, moreover, may be used to develop tunable tinted center lasers.

In this regard the purpose of the present study was to carry out comprehensive investigations of the photoexcitation processes of nonequilibrium carriers in wideband dielectrics under the influence of nanosecond UV laser pulses by examining photoconductivity, the drag effect and nonlinear absorption.

The primary test materials were undoped, alkali-halide crystals without any deliberate impurities (NaCl, KCl, KBr, KI, CsI, CsBr, RbCl) as well as KDP, DKDP, ruby and fluorite crystals. These samples were selected since the bandgap of the energy states for these crystals varies over a significant range (6-9 eV) which makes it possible to follow the photoexcitation properties in the UV and from the practical viewpoint by the fact that such crystals are widely employed in laser optics: Alkali-halide crystals are widely used as passive (modulators, windows, etc.) and active (tinted center lasers) elements in laser systems, while KDP and DKSP crystals are used as nonlinear optical frequency converters and ruby and fluorite are used as active laser elements.

Today the majority of experimental investigations of the phenomenon of photoconductivity employ semiconductor materials, although historically alkali-halide compounds were the first test specimens, since the method of growing semiconductor monocrystals in the 1920s and 1930s had not yet been developed, and the use of test substances in the form of powders caused their photoconductivity to significantly influence the various contact phenomena [21].

Investigating photoconductivity in transparent materials, particularly in wideband crystals, presents significant experimental difficulties. The virtual absence of dark conductivity makes it difficult to employ the standard photocurrent measurements typical of semiconductors employing ohmic contacts. The small values of the multiphoton absorption coefficients, and the deep-lying ionized defects in the bandgap produce very low photoresponse. Therefore the role of interfering, parasitic effects such as carrier injection from the electrodes or the influence of surface photoconductivity increases. Clearly this explains the relatively small volume of studies devoted to the photoconductivity of wideband dielectrics.

In the present study significant attention was devoted to the experimental technique and the development of methods of investigating the photoconductivity, the drag effect and nonlinear absorption. A comprehensive investigation of such phenomena required the development of a high-power single-mode laser that would make it possible to employ emission at 1.06, 0.53, 0.35 and 0.27 μm in a timely fashion.

Chapter 1 of the present study presents a brief survey of research into the photoconductivity of dielectric crystals. Primary attention was focused on an analysis of studies devoted to laser-excited photoconductivity (henceforth we will occasionally use the term "laser photoconductivity"). We note that the published results from experimental investigations of the photoconductivity of dielectrics are often contradictory and therefore do not yield a specific answer to the question of the origin of the nonequilibrium carrier photoexcitation mechanisms. The primary conclusions deriving from studies [22-25] are investigated in detail; these studies investigated the dependence of the photoconductivity of undoped (i.e., not containing any deliberate impurities) alkali-halide crystals on the intensity of the laser excitation emission.

Chapter 2 contains a description of the assembly used in experiments on laser photoconductivity of dielectric crystals in the UV. Attention is focused on the problems of proper techniques for recording photoconductivity signals from the test sample bulk. Optical and electrical diagrams of the assembly are given together with its primary specifications. The selection of wideband dielectric samples used as the test samples is substantiated.

Chapter 3 presents results from an experimental investigation of photoconductivity carried out over a wide range of laser excitation intensities ($I \sim 10^3 - 10^8$ W/cm^2) at 0.27 and 0.35 μm. Specifically it is established that in alkali-halide crystals the intensity dependence of photoconductivity is quite complex, jumping from linear to near-quadratic at $I \sim 10^4 - 10^5$ W/cm^2 and back again to linear at $I \sim 10^6 - 10^7$ W/cm^2. An anomalous hysteresis nature of this relation was also discovered that resulted in an irreversible increase in photoconductivity. A qualitative interpretation of the observed phenomena is given.

Data are given on the photoconductivity of KDP and DKDP crystals indicating the realization of two-photon carrier excitation processes in such crystals (at $I \geqslant 5 \cdot 10^6$ W/cm^2). A drop in photoconductivity with multiple UV excitation of constant intensity was discovered in ruby and was attributed to the formation of space charge which makes it possible to estimate the specific conductivity in dielectric crystals with a sufficiently long Maxwell relaxation time ($\geqslant 10^2$ sec). At high UV laser intensities ($I \sim 10^8$ W/cm^2) photoconductivity of the fluorite crystals was recorded and was attributed to ionization of the impurity states.

Chapter 4 presents results from experimental investigations of the photon drag effect of carriers in an electromagnetic wave field and the nonlinear absorption of laser emission energy (using the optoacoustic method) in alkali-halide crystals in the UV. These results in conjunction with experimental data on the photoconductivity of identical crystals made it possible to reliably substantiate the conclusion that at high laser emission intensities ($I \geqslant 10^7$ W/cm^2) the dominant photoexcitation process is two-photon carrier creation in the presence of quadratic carrier recombination. The data presented in this chapter made it possible to estimate the lifetime of the nonequilibrium carriers as well as the value of the two-photon absorption coefficient in the test crystals.

Chapter 5 gives a theoretical interpretation of the photoconductivity regularities observed in alkali-halide crystals based on a kinetic model that includes single-photon ionization of impurities, two-photon ionization of the crystalline lattice atoms and linear and quadratic carrier recombination. The model accurately explains the entire set of experimental data. The irreversible increase in photoconductivity is attributed to the radiation creation of defects associated with the process of quadratic electron-hole recombination.

CHAPTER 1

A BRIEF SURVEY OF STUDIES ON THE PHOTOCONDUCTIVITY OF DIELECTRICS

The present chapter does not proport to provide a comprehensive analysis of existing studies devoted to the photoconductivity of dielectric crystals, since a significant number of publications are devoted to this topic. Primary attention was devoted to research on photoconductivity excited by laser emission in undoped monocrystals without deliberate introduction of impurities. Research results in which the dependencies of photoconductivity on emission intensity were investigated in greatest detail.

1. Early Studies on the Photoconductivity of Dielectrics

The first adequate description of photoconductivity was provided as early as the 1920s by Gudden and Pohl who experimentally investigated zinc sulfide, diamond and alkali-halide compounds (a sufficiently comprehensive survey of their early studies devoted to this topic may be found in [16]). Specifically they established the quantum nature of the phenomenon and demonstrated that the existence of impurities expands the range of photoconductivity to longer wavelengths and that the charge remaining in the crystal after excitation either partially or totally disappears upon crystal heating.

In analyzing early studies on the photoconductivity of dielectrics it may be noted that by the end of the 1930s the relationship between photoconductivity and principally the volumetric properties of various substances had been established reliably and that the charge carriers in this case were free electrons [16].

It is interesting to note that in investigating the photoconductivity of dielectrics the most common samples are crystals with high (of the order 10^{18} - 10^{20} cm^{-3}) concentrations of impurities deliberately added to the samples during the growth process (doped crystals) or crystals irradiated by high energy radiation in order to generate a high defect concentration (X-rayed crystals). Investigations of the photoconductivity of such crystals normally focus on the influence of emission with a quantum energy close to that of the impurity or defect levels. Therefore the purpose of such research efforts normally involve determining the structure of the impurity or defect energy levels.

Therefore [26] has investigated the photoconductivity of nitrogen-containing diamonds in the longwave portion of the visible spectrum and the fine structure in the spectral distribution of the photocurrent within the absorption band was investigated. The impurity nitrogen concentration was rather high: Approximately 10^{20} cm^{-3}. It is interesting to note that the relative level and clarity of the minima and maxima of the photocurrent on its spectral distribution curve in [26] was different not only for different samples but also for different sites in the same sample, which indicates that the impurities existing in the crystal make a significant contribution to the photoconductivity of doped dielectrics as well as their volumetric distribution.

In [27] the diamond was doped by lithium ions in a concentration of 10^{15} - 10^{16} cm^{-3} and the microwave technique at room temperature was used to record photoconductivity. The spectral distribution of the photoconductivity of the samples in the 3-13 μm range was also measured.

Studies [28] investigated the absorption spectra and photoconductivity of alkali-halide crystals containing colloidal potassium particles; this study employed KBr crystals while [29] investigated KCl crystals. The peak of the photoconductivity signal was located in the visible range at approximately 0.6 μm.

The photoconductivity of electrolytically-tinted NaCl crystals was investigated in [30] where a carrier lifetime of $\sim 4 \cdot 10^{-4}$ sec was established from an analysis of the spectral distribution (in the visible range) and the temporal decay of the photocurrent; the sample polarization was also determined, i.e., the direction of photocurrent changed after the excitation light was switched off.

The authors [31] observed photoconductivity in alkali-halide melt-grown crystals containing hydroxide impurities of alkaline metals and established that unlike "normally pure" alkali-halide crystals, photocurrent and destruction of the crystal tinting were observed in X-ray irradiated KCl-KOH and KBr-KOH crystals at low temperatures, in the order of 100 K, which corresponds to a sharp drop in the concentration of F-centers.

Thus, [31] discovered a correlation between the optical damage of F-centers and photoconductivity. Its cause, according to the authors of [31], is the reduction in the activation energy of the F-centers compared to "pure" crystals due to the existence of anion impurities. This study also compared the photoconductivity of pure and impurity KCl and KBr crystals. With an identical concentration of F-centers the photoconductivity of melt-grown alkali-halide crystals containing the alkaline metal hydroxide impurity was approximately an order of magnitude lower, which was attributed by the authors to a reduction in carrier mobility due to the anion sublattice.

We should point out here that in [26-31] the emission sources employed to excite photoconductivity in the doped wideband dielectrics were not laser sources and consequently short pulses and significant intensities were not used to irradiate the material which in turn made it impossible to investigate the nonequilibrium carriers with characteristic times shorter than 10^{-6} sec (and specifically made it impossible to determine such an important parameter as the nonequilibrium carrier lifetime), or to record the photoconductivity signals over a broad range of excitation emission intensities. This in fact made it impossible to investigate the photoconductivity of undoped dielectric monocrystals.

Moreover, only by employing high intensity laser emission to irradiate dielectric crystals is it possible to study multiphoton energy absorption processes as well as the various phenomena and effects that have been observed previously in semiconductor crystals (such as the Dember and Hall effects, the photon drag of free carriers, etc.).

Moreover, the interaction of high intensity laser emission with broadband dielectrics is required to observe specific processes occurring at high excitation energies: The formation of nonequilibrium F-centers and other radiation defects.

2. Laser Photoconductivity of Dielectrics

The development of laser technology accelerated the investigation of processes occurring from laser emission action on dielectrics. As in the case of nonlaser excitation sources the majority of works devoted to photoconductivity arising in dielectrics under laser emission action employed doped crystals as the test samples.

Thus, for example, study [32] at the ruby laser wavelength of $\lambda = 0.69$ μm observed photoconductivity of Eu^{+2} ion doped KCl crystals while in [33] photoconductivity of Tl^{+} ion doped alkali-halide crystals was observed at the same wavelength; the KCl crystals investigated in [34] contained a significant Ag^{+} ion impurity. The concentration of impurity ions was rather high and amounted to 10^{16} - 10^{19} cm^{-3}. A detailed description and analysis of the experiments conducted by the authors of studies [32-34] may be found in survey study [35] which specifically derived the dependencies of the photoluminescence intensities of impurity alkali-halide crystals on the photocurrent level: With a power of 4 at 0.69 μm and of 6 at 1.06 μm.

In pure KCl samples used by the authors of [35] for comparison, similar multiphoton excitation was observed at laser emission intensities two orders of magnitude greater than the intensities yielding an analogous effect in crystals containing a deliberate impurity. One interesting fact is that [35] employed non-Q-switched lasers (the laser pulse duration was 0.5-2 msec), while photoresponse was observed only at very high pulse energies of ~ 1 J which suggests a strong influence of contact phenomena and surface photoconductivity on the test sample.

Study [36] determined that the excitation of carriers in thallium-, indium- and europium-activated alkali-halide crystals when exposed to nitrogen laser emission in the longwave impurity absorption band (0.337 μm) produces phosphorescent, photoconductivity and thermally-stimulated luminescence.

An analysis of the temperature dependence of the photocurrent led the authors of [36] to the conclusion that photoconductivity was excited by single-photon excitation. Therefore there is clear discrepancy in the conclusions reached by the authors of studies [35] and [36] with respect to the role of multiphoton processes in the observed phenomena.

Results from recent research [25, 37-41] have revealed that in order to investigate the laser photoconductivity of wideband dielectrics it is necessary to use nanosecond or picosecond pulses. Moreover the more interesting important aspects from the viewpoint of investigating intrinsic (characteristic of only a specific given wideband dielectric) processes of carrier excitation by laser emission include investigations of photoexcitation of relatively pure crystals that do not contain deliberately introduced impurities.

Studies [17, 18, 22, 23] are among some of the first studies devoted to the laser photoconductivity of undoped wideband dielectrics. Study [22], for example, investigated untinted NaCl and Al_2O_3 monocrystals at the wavelength of a ruby laser employing a laser pulse duration of 10^{-8} sec at an intensity ~ 100 MW/cm^2. Power dependencies of the charge induced across the electrodes on the laser emission excitation intensity were observed: The power was 5 for NaCl and 3 for Al_2O_3; the observed photocurrent rise times were ~ 2 msec. The authors of [22] interpreted their results as multiphoton energy absorption.

In [23] in free-running lasing of ruby and neodymium lasers (pulse duration 0.5-2 msec) photoconductivity of untinted alkali-halide crystals was observed together with its dependence on the excitation laser emission intensity with a power between 4 and 6 which the authors interpreted as multiphoton excitation of the exciton states or band-to-band transitions.

The interpretations of the authors of [22, 23] do, however, raise serious questions for the following reasons. First multiphoton ionization processes with such a large number of quanta absorbed in the single process have been reliably observed to date only in gases (see [42, 43]), so without special and careful substantion the observed dependencies of the photoconductivity signal on the laser emission excitation intensity cannot be attributed to p-photon absorption (for p = 3-6).

Second the intensity dependences of photoconductivity with a power between 3-6 observed in [22, 23] are in good agreement with the surface photoconductivity phenomenon [18, 39, 40, 44] which arises if the external electric field near the irradiated crystal surfaces is of the same order as in the crystal bulk. The characteristic lifetime of the nonequilibrium carriers near the surface (if it has not been specially treated) is several orders of magnitude greater [40] than in the bulk, while the signal amplitude is highly dependent on the laser emission power density: The power of the dependence of the photoconductivity on laser emission intensity is determined by the difference in the energies of neighboring pulses and in experiment virtually any values may be achieved [18, 40].

Therefore without taking special preventative measures which are not noted in [22, 23] experiments to investigate laser photoconductivity of dielectrics will reveal a photoresponse generated by the excitation of surface states whose amplitude will, moreover, be several orders of magnitude greater than the photoresponse from the crystal bulk [39].

Third, it is quite probable that the high-power dependencies observed by the authors of studies [22, 23] characteristic of multiphoton carrier excitation processes may also be attributed to carrier injection by the photoeffect from the electrodes (unfortunately the electrode material is not noted in [22, 23]), since we have observed precisely such high-power dependencies of the photoconductivity signal on the excitation laser emission intensity in developing the experimental technique when nickel, silver and copper electrodes were employed [41].

And, finally, fourth, with the photoflux values of $Q \simeq 3 \cdot 10^{22}$ $cm^{-2} \cdot$ $\cdot$ sec^{-1} employed in [23] the laser emission intensities with a quantum energy $\hbar\omega$ = 1.78 eV (ω is the emission frequency) are $\sim 10^4$ W/cm^2 (in [22] they come out to 10^8 W/cm^2). Assuming that the sensitivity of the technique made it possible to measure nonequilibrium carrier concentrations of better than 10^8 cm^{-3} (see [25, 41]), we will estimate the four-photon absorption coefficient in the experiments conducted by the authors of [23]. The rate of carrier generation with a recombination time $\sim 10^{-8}$ sec [25] is $dn/dt \sim 10^{16}$ $cm^{-3} \cdot sec^{-1}$.

Since $dn/dt \sim \text{æ}Q/4$, where æ is the energy absorption coefficient, $Q/4 \sim 10^{22}$ $cm^{-2} \cdot sec^{-1}$ is the photoflux in four-photon energy absorption, $\text{æ} \sim 10^{-6}$ cm^{-1}.

In the obvious expression for multiphoton energy absorption

$$\text{æ} = \text{æ}_1 + \text{æ}_2 I + \text{æ}_3 I^2 + \ldots, \quad (1)$$

where æ_1 is the linear absorption coefficient, æ_2 is the two-photon absorption coefficient, etc., and I is the laser emission intensity, the $\text{æ}_4 I^3$ term characterizes absorption in the four-photon process. If, like the authors of [22, 23] we take energy absorption to be primarily multiphoton absorption (in this case four-photon absorption), we obtain for $I = 10^4$ W/cm^2 $\text{æ}_4 \sim 10^{18}$ $W^{-3} \cdot cm^5$.

Such a four-photon absorption coefficient means that at laser emission intensities greater than 10^8 W/cm^2 the absorption coefficient will exceed 10^6 cm^{-1}. This in turn suggests that at these emission intensities the dielectric crystal will generally become nontransparent. Based on recent research on laser-induced damage we know, however, that alkali-halide crystals are transparent at the wavelength of ruby and neodymium lasers and are not damaged in the crystal bulk up to significantly high emission intensities of the order 10^{11} W/cm^2 (see, for example, [7]).

This entire discussion makes it possible to conclude that in [22, 23] and in the later study [24] multiphoton energy absorption processes in solids were in fact not realized, but rather the observed high power dependencies of the photoconductivity signal on the laser emission intensity may more accurately be attributed to surface photoconductivity and the photoelectric electrode effect.

Studies [17, 18] discovered laser photoconductivity of ruby at room temperature at 0.69 μm and at laser emission intensities of $\sim 10^{10}$ W/cm^2 (i.e., near the damage threshold of ruby and in a very narrow intensity range, which made it impossible to measure the dependence of photoconductivity on the intensity of the incident laser emission) in Q-switched operation. The laser pulse duration was $5 \cdot 10^{-8}$ sec. The measured ruby photoconductivity pulse duration was $9 \cdot 10^{-7}$ sec which could be attributed to electron trapping by the minor traps whose depth did not exceed kT (k is Boltzmann's constant, T is absolute temperature).

The authors of [45, 46] measured photoconductivity in lithium niobate crystals. Study [46] proposed a method of detecting very low photoconductivity signals by investigating the external electric field screening kinetics which in bifringent crystals will change the degree of bifringent in the laser-irradiated region, which is easily registered by the polarization method. The study employed a CW He-Ne laser (0.63 μm).

For pure lithium niobate at laser emission intensities ~ 20 W/cm^2 the photoconductivity was $3 \cdot 10^{-14}$ Ohms^{-1} $\cdot$ cm^{-1}. In the case of iron doping the photoconductivity value increased by two orders of magnitude, which, in the authors' opinion, is quite natural since electrons from the impurity energy levels make a contribution to the signal.

For undoped samples of the same crystals the photoconductivity dependence on the excitation laser emission intensity and on temperature was measured: With intensity changes up to 10^3 W/cm^2 the photoconductivity followed a linear progression; with a sample temperature change from 20 to 150°C the photoconductivity increased exponentially.

We emphasize that the method of recording photoconductivity described in [46] is applicable only to bifringent crystals.

The lifetime of electron photoconductivity at room temperature was measured in the experiments of [24] on induced multiphoton conductivity of alkali-halide crystals not containing specially-introduced impurities at 0.69 and 0.347 μm (laser pulse duration of 20 nsec), which was interpreted by the authors as the nonequilibrium carrier lifetime, and was ~ 1 μsec.

Study [24] obtained the dependencies of the charge q induced across the electrodes for KI crystals on the excitation laser emission intensity: $q \sim I^4$ at $\lambda = 0.69$ μm and $q \sim I^2$ at $\lambda = 0.347$ μm. For NaCl and KCl crystals $q \sim I^5$ at $\lambda = 0.69$ μm and $q \sim I^3$ at $\lambda = 0.347$ μm.

We immediately see a discrepancy in the experimentally-determined non-equilibrium carrier lifetimes in the alkali-halide crystals obtained by the authors of [24] and the results of authors of other studies (see [22, 25, 38]). For example in [25] at the same second harmonic of a ruby laser (0.347 μm) at analogous laser emission intensities, two-photon energy absorption was observed in a KBr crystal, while the measured nonequilibrium carrier lifetime was $\sim 10^{-11} - 10^{-9}$ sec depending on their concentration.

It is likely that the authors of [24] did not take measures to help eliminate the influence of surface photoconductivity and the photoelectric effect from the electrodes on the measures signal, as discussed above in connection with [22, 23].

Moreover the laser employed in [24] was evidently not a single-mode laser which would significantly distort the experimental results on laser photoconductivity, since in this case the emission intensity in the beam will vary from pulse to pulse quite significantly and in an unpredictable manner.

The nonequilibrium carrier lifetime in undoped alkali-halide crystals was measured experimentally in [25] at 0.347 μm at a laser pulse duration of 20 psec by investigating the photoresponse. Such a short picosecond pulse duration made it possible to also measure the shorter carrier lifetimes by using laser emission intensities below the breakdown thresholds of the test materials.

At low excited carrier densities ($\sim 10^{12}$ cm^{-3}) carrier lifetimes of the order 10^{-9} sec were detected in all test alkali-halide crystals which is attributed by the authors to the existence of shallow impurity states in the bandgap. At higher excited carrier concentrations ($\geqslant 10^{13} - 10^{14}$ cm^{-3}) their lifetime was significantly shorter (of the order 10^{-11} sec), while the carrier lifetime dropped progressively with growth of the carrier density. The authors of [25] attributed their result to quadratic recombination of the nonequilibrium carriers at sufficient carrier densities.

These results are in good agreement with previous data from an investigation of defect formation in alkali-halide crystals obtained in [37, 38, 47] which demonstrated experimentally that at free carrier densities significantly less than densities achieved at the optical breakdown thresholds (several orders of magnitude), the carrier lifetime is shorter than 10^{-11} sec.

Studies [39, 40] investigated photoelectric phenomena in KH_2PO_4 (KDP) and $KDHPO_4$ (DKDP) crystals at room temperature at the neodymium laser wavelength of 1.06 μm and at its second harmonic (0.53 μm). The laser pulse duration was $4 \cdot 10^{-8}$ and $3 \cdot 10^{-8}$ sec, respectively.

According to the authors at 1.06 μm the volumetric photoresponse could be attributed to the nonstationary heating of the crystalline lattice and the change in the low-frequency dielectric constant, since the absorption coefficient of KDP at this wavelength is better than $5 \cdot 10^{-2}$ cm^{-1} [40].

At 0.53 μm the observed photoresponse was attributed to the generation of nonequilibrium carriers, while the photocurrent amplitude was linearally dependent on the laser emission intensity, which led the authors to interpret this as single-photon ionization of the impurity states (the energy absorption coefficient in KDP and DKDP at 0.53 μm was assumed to be less than 10^{-4} cm^{-1}).

Studies [39, 40] also measure the lifetime of nonequilibrium carriers in KDP ($\leqslant 3 \cdot 10^{-8}$ sec) and estimated their maximum (in experimental conditions at $I \sim 10^8$ W/cm^2) concentration of $n \sim 10^{12}$ cm^{-3}. In reality this means that the nonequilibrium carrier lifetime itself was not measured but rather the laser photoconductivity lifetime equal to the laser pulse duration. Hence we may conclude that the results from [39, 40] aimed at determining carrier lifetime do not contradict results from [25, 37, 38] in the sense that they provide only the upper bound of the measured quantity.

An analysis of these studies reveals that the phenomenon of laser photoconductivity may be successfully employed as a sensitive method (see, for example, [48]) of recording many processes that occur in the interaction of high-power electromagnetic emission with matter. Recent experimental studies [25, 37, 38, 47] indicate that in pure (without deliberately introduced impurities) alkali-halide crystals the nonequilibrium carrier lifetime at room temperature does not exceed 10^{-11} sec, if the carrier concentration is $n \geqslant 10^{12} - 10^{13}$ cm^{-3} and less than 10^{-8} sec for $n \leqslant 10^{10}$ cm^{-3}.

Virtually all authors have noted that in wideband dielectrics the photoconductivity signal is linearally dependent on the high-voltage applied to the sample. Special introduction of impurities (doping) and the deliberate creation of defects in the crystal bulk (by X-ray irradiation), as well as thermal annealing of the samples will increase the nonequilibrium carrier lifetime, while chemical treatments of the crystals will reduce this parameter [25].

In addition the dependencies of the charge induced across the electrodes which is proportional to the nonequilibrium carrier concentration on the laser emission excitation intensity were highly divergent among the authors. In many cases (see [22-24]) it was difficult to attribute these relations to anything other than the influence of surface photoconductivity on the signal and the electrode photoeffect. The role of multiphoton carrier excitation in alkali-halide crystals in investigating their photoconductivity is therefore not clear. Moreover we should note that in the studies we analyzed in which the dependencies of the photoconductivity of the dielectric crystals on the excitation laser emission intensity were obtained the range of variation in the intensities was, as a rule, quite small which in fact made it impossible to sufficiently investigate the carrier photoexcitation processes.

We know of only two studies [49, 50] from one team of authors devoted to an experimental investigation of the two-photon energy absorption processes in wideband dielectrics which measured the absolute two-photon energy absorption coefficients at 0.335 and 0.266 μm in the UV. The authors indicate that in materials with a bandgap greater than $2\hbar\omega$ no noticeable nonlinear energy absorption was observed up to intensities corresponding to the breakdown thresholds of the sample surfaces, i.e., $I \sim 10^8$ W/cm^2. In these studies the two-photon energy absorption coefficients were measured by the absorption of light transmitted through a test sample.

It should be noted that one of the first theoretical analyses of multiphoton ionization processes in solids was carried out in [51] which makes it possible to perform certain important estimates (such as the values of the multiphoton absorption coefficients, their dependence on the effective electron mass and on the crystal bandgap) and also stimulated further theoretical investigations on this topic. The theory developed in [51] specifically predicted a frequency dependence of the two-photon absorption coefficient which has been accurately confirmed by experiment [49].

CHAPTER 2

EXPERIMENTAL SET-UP, CONDITIONS AND TECHNIQUES FOR LASER PHOTOCONDUCTIVITY MEASUREMENTS

In this chapter we will examine the issues associated with the development of the laser assembly used to investigate the photoconductivity of dielectrics in the UV. In order to conduct a comprehensive investigation of laser photoconductivity it is very important that a multipurpose assembly be developed, i.e., one that makes it possible to use laser emission over a broad range of intensities and at different wavelengths both in isolation and in any combinations necessary for experimental applications. Thus (for example as noted in [6, 7]) the modal composition of the emission is an important aspect.

The authors of the majority of studies devoted to the laser photoconductivity of dielectrics have not given sufficient attention to the selection of the material from which the electrodes are fabricated (with a capacitive legitimate signal recording circuit) and in a number of cases did not indicate the material selection. Such an approach cannot be considered methodologically sound, since the irradiation of the electrodes by direct or scattered laser emission, particularly emission with a high quantum energy, may cause carrier injection from the electrodes to the test sample and therefore will distort the photoconductivity signal. It is also necessary to focus on measures aimed at eliminating the contribution of surface conductivity to the recorded signal which may significantly alter the measured relations [39, 40].

At present there exists only a few experimental studies on the laser photoconductivity of undoped wideband dielectrics. Investigating specifically undoped monocrystals will make it possible to establish the fundamental physical regularities associated with the intrinsic photo-ionization mechanisms in the matrix. It was this fact that was responsible for our choice of alkali-halide crystals as the test samples.

Alkali-halide crystals are transparent at all our wavelengths and have a rather simple cubic structure, while their physical properties are rather well known. Extensive theoretical and experimental studies, particularly recently, have been devoted to the formation and investigation of defects in alkali-halide crystals (see [37, 38, 47]). Moreover alkali-halide crystals represent one of the primary model test samples for investigating the maximum optical laser radiation strength of matter.

Nonetheless many issues associated with the fundamental properties of matter remain unresolved for the majority of wideband dielectrics (including alkali-halide crystals). Specifically there is insufficient experimental data on the properties and structure of the defect and impurity energy levels and few reliable results on multiphoton absorption processes. For example prior to 1972 the literature contained no correct information on the lifetimes or concentrations of nonequilibrium carriers.

Therefore currently available data make it possible to state that in undoped alkali-halide crystals the mechanisms responsible for laser emission-excited photoconductivity have not been unambiguously established.

In addition to alkali-halide crystals we investigated in the UV the laser photoconductivity of potassium dihydrophosphate (KDP), potassium dideuterophosphate (DKDP), fluorite and ruby crystals. As we know these crystals are widely used in various nonlinear optical devices and in high-power laser systems in which the problem of the maximum laser radiation strength of matter is still critical and therefore new data on the properties of such substances resulting from their interaction with high-power laser emission are quite important.

3. Laser Assembly

In the experiments we employed a single-mode Q-switched YAG:Nd^{3+} laser. In order to investigate the interaction processes of high-power laser emission with matter it was methodologically important to employ single-mode lasers, since this allowed exact measurement of light beam intensities as well as proper comparison of experimental results obtained by different authors and at different wavelengths. Moreover, the nonreproducibility of the modal structure of the laser emission when using multimode lasers for the research makes it impossible to perform correct comparisons of experimental results to the conclusions and estimates derived from theory.

The master oscillator of the laser assembly employed a YAG crystal 7 mm in diameter and 120 mm in length in the laster oscillator. The lateral surface of the crystal was frosted and a bleaching coating at 1.06 μm was applied to the crystal ends. A single-lamp illuminator employing an IFP-1200 flashlamp was used as the pump source.

The optical cavity was formed by a 100% mirror and a three-layer silica stop which served as the outcoupling mirror and was designed for longitudinal mode selection. The cavity contained a passive shutter containing a solid-polymer-based saturable dye (the cavity Q-switch) together with a 1.5 mm ϕ diaphragm which was the transverse mode selector for extracting the fundamental TEM_{00} transverse mode. The configuration of the passive shutter at the Brewster angle to the optical axis of the laser caused the emission to be linearally polarized.

In order to obtain sufficient power for the investigations the master oscillator emission was amplified in three single-pass YAG:Nd^{3+} crystals 10 mm in diameter and 120 mm in length. Depending on the experimental processes a somewhat different configuration of the amplification stage is possible where neodymium glass 300 mm in length and 15 mm in diameter is used as the third stage.

In order to avoid self-excitation the entrance end of the garnet crystals was bleached while the exit end was bevelled at an angle of 1° with respect to the entrance end. Single-lamp illuminators employing IFP-1200 flashlamps were used to pump the garnet crystals in the amplifier stages of the laser, while a two-lamp illuminator employing IFP-5000 flashlamps was used to pump the neodymium glass.

The output laser emission energy at 1.06 μm in such a system was 1 J for a pulse duration of 10 nsec at half-amplitude. A FEK-19 photocell and a I2-7 high-speed broadband oscilloscope were used to measure the laser pulse duration and monitor its temporal waveform. This monitoring system provided a bandwidth of the order 3 GHz.

The laser emission in the experiments was single-frequency emission as verified by a Fabry-Perot interferometer. The laser pulse had a near-Gaussian waveform.

Fundamental emission at 1.06 μm was converted (depending on the experimental problem) into the second, third and fourth harmonics. After amplification the fundamental emission was guided to a nonlinear KDP crystal where interaction produced the second harmonic at 0.53 μm. The conversion efficiency was 20%.

The second harmonic emission was then either directed to a KDP crystal where the first and second harmonics were added forming the third harmonic (at 0.35 μm) or was fed to a doubling DKDP crystal in which 90-degree wave synchronism was achieved at a temperature of 38.7°C from *ooe* interaction and the fourth harmonic was formed at 0.27 μm. The temperature stabilization accuracy of 0.05°C for the DKDP crystal was maintained by a special electronic circuit.

The maximum laser emission energy at frequencies in the UV (0.35 and 0.27 μm) was 0.1 and 0.04 J respectively for a laser pulse 8 nsec in duration at half-amplitude. The signal waveform was near-Gaussian.

The energy output was monitored by a high-sensitivity high-speed pyroelectric photodetector based on polycrystalline organic substances [52] by extracting part of the emission by means of a plane-parallel silica plate. Such a system is highly linear across the entire sensitivity range and produces no electrical static in the measurement circuits. A IMO-2N calorimeter was used to calibrate the pyrodetector.

When necessary the laser emission was attenuated in steps by means of calibrated HC glass filters of variable optical density at 0.35 μm and UFS-1, UFS-2, BS-3, BS-12 glasses for 0.27 μm. The transmission linearity of the light filters was monitored across the entire test range of laser emission intensities. The power was continuously varied by altering the flashlamp pump energy in the laser amplifier stages.

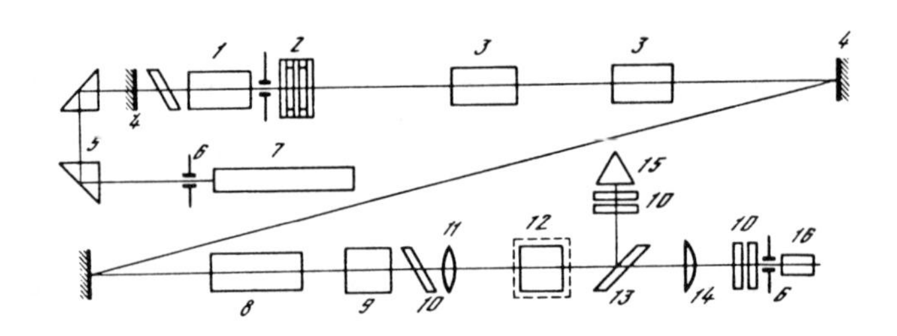

Figure 1. General experimental set-up to investigate the photoconductivity of wideband dielectrics.

1 - YAG:Nd^{3+} laser; 2 - silica stop; 3 - YAG:Nd^{3+} laser amplifiers; 4 - mirrors; 5 - prisms; 6 - diaphragms; 7 - tunable He-Ne laser; 8 - glass-Nd^{3+} amplifier; 9 - KDP crystal doubler; 10 - light filters; 11 - lens, $f_0 \sim 3$ m; 12 - DKDP second harmonic doubler; 13 - beam-splitting silica plate; 14 - silica lens, $f_0 \sim 500$ mm; 15 - pyrodetector for energy control; 16 - test sample.

The laser beam diameter was between 0.2-4 mm when collimated or weakly-focused by a long-focus KU fused silica lens (focal distance $f_0 \sim 500$ mm) at the entrance to the test samples measured at $1/e$ of the maximum intensity on the axis. When necessary broader beams could also be used by placing the diagram in immediate proximity to the test dielectric sample. This made it possible to measure the photoresponse over a wide range of laser emission intensities: From values limited by the maximum sensitivity of the recording equipment ($\sim 10^3$ W/cm^2), to values near the optical breakdown threshold of the crystal surfaces ($\sim 10^8 - 10^9$ W/cm^2) [53].

The general configuration of the laser set-up is shown in Figure 1.

4. Measurement Assembly For Photoconductivity Measurements

In measuring the laser photoconductivity we employed the signal recording configuration developed on the basis of general circuit principles with capacitive coupling between the electrodes and the sample (see, for example, [54]). In our case the photoconductivity signal recording circuit is shown in Figure 2 (see also [41, 55]).

High-voltage generator G based on a high-voltage pulsed transformer and a rectifier in addition to firing the laser pump lamps also generated a 12 kV electric pulse with a duration of near $2 \cdot 10^{-3}$ sec. Such a generator operating mode made it possible to avoid space charge generation in the test dielectric samples over the time period between laser pulses amounting to 10-15 sec, since the majority of our test materials had a rather high ion conductivity value at room temperature (for example, on the order of 10^{-13} $\text{Ohms}^{-1} \cdot \text{cm}^{-1}$ for KDP crystals [56] or on the order of 10^{-11} $\text{Ohms}^{-1} \cdot \text{cm}^{-1}$ for NaCl crystals) and a correspondingly short dielectric relaxation time ($\leqslant 1$ sec). For processes occurring in time periods comparable to the duration of a laser pulse ($\sim 10^{-8}$ sec) the high voltage applied to the sample may be assumed to be d.c. voltage.

By the time the laser giant pulse is generated the voltage across capacitor C_o (see Figure 2) will have reached a maximum, as determined by the selection of the resistance values of R_o, R_1 and the capacitance of C_o. The photoconductivity signal of the sample was taken from load resistor R_H and was either directly input to a C8-2 storage oscilloscope or was fed through a low-noise broadband signal amplifier with a low input capacitance.

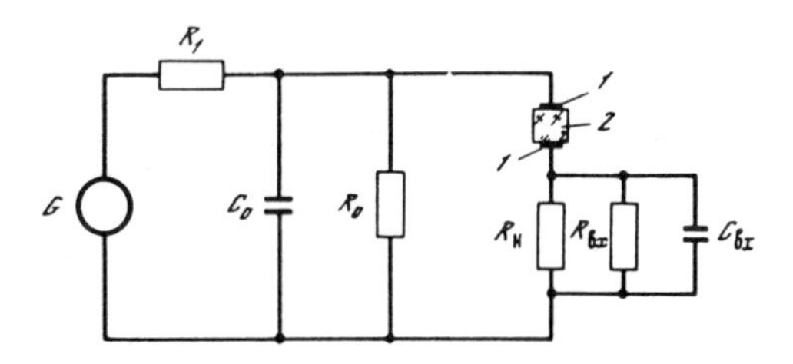

Figure 2. Recording circuit for recording the photoconductivity signal.

G - Pulsed high-voltage generator; R_H - load resistor; R_1, R_o, C_o - resistors and capacitors for generating a high-voltage pulse ~ 1 msec in duration; R_{BX}, C_{BX} - input resistance and capacitance of the measurement circuits; 1 - platinum electrodes; 2 - test sample.

The photoconductivity recording circuit made it possible to investigate the photoresponses both with integration of the legitimate signal and with high temporal resolution. If the conditions $R_H \ll R_{BX}$ and $R_H C_{BX} \gg \tau_И$ are satisfied where R_{BX}, C_{BX} are the input resistances and capacitance of the oscilloscope or the broadband amplifier, and $\tau_И$ is the laser pulse duration, integration of the photoconductivity signal is achieved which makes it possible to improve the sensitivity of the recording circuit to low amplitude signals, i.e., to expand the dynamic range of laser photoconductivity investigations. In this case the signal has a sharp leading edge and an exponential decay over $t = R_H C_{BX}$. The amplitude of the photoconductivity signal may be represented as

$$u = U\tau_и/(R\ C_{вх}), \tag{2}$$

(evidently in our cases the carrier recombination time is shorter than the pulse duration [25, 37, 38] and therefore $\tau_И$ is included in formula (2) for determining the lifetime of laser photoconductivity), where R is the resistance of the sample during the laser pulse, U is the voltage applied to the sample, and $U \gg u$.

The conductance of the sample σ is determined by the resistance R as well as the geometric dimensions of the sample and the electrodes:

$$\sigma = f/(Ral), \tag{3}$$

where f is the distance between electrodes, a is the electrode width, l is the electrode length (for now we assume that the electrode width is equal to the width of the test dielectric sample in the cross-section to the laser beam).

Formulae (2) and (3) are written for the case where the test sample is entirely "filled" with the laser emission. In actuality the equivalent electric circuit of the sample may be represented as the capacitances of unilluminated and illuminated sections of the sample, while the conductance is "connected" in parallel to the latter. In this case it is easy to demonstate that the voltage across the load resistance is determined by the formula

$$u = K_з U\tau_и/(R\ C_{вх}), \tag{4}$$

where K_3 is the fill factor of the interelectrode gap by laser emission (Figure 3): $K_3 = S/(af)$ (S is the cross-sectional area of the laser beam). Thus, accounting for the fill factor the amplitude of the photoconductivity signal may finally be written as

$$u = U\tau_и \sigma l S/(f^2 C_{вх}). \tag{5}$$

When the inequality $R_H C_{BX} \ll \tau_И$ is satisfied, the signal recording circuit operates with a high temporal resolution making it possible to analyze a photoresponse pulse waveform with a duration of $\sim 10^{-9}$ sec. The photoconductivity signal amplitude in this case will be

(6)

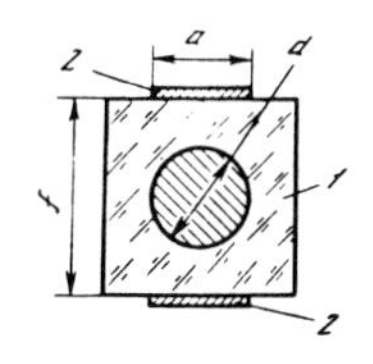

Figure 3. Laser emission irradiation of the test sample.

1 - Sample (in cross-section); 2 - olatinum electrodes: a - width; f - interelectrode distance; d - diameter of the collimated laser beam. The hatched area indicates area filled with laser radiation.

5. Photoconductivity Measurement Method

The test dielectric samples transparent to laser emission were crystals in parallelepiped form 4 x 4 mm^2 in cross-section and $l_{обр} \sim 20$ mm in length with natural shear along the cleave planes or with polished surfaces. Such characteristic sample dimensions were near-optimum for our applications, since the extraordinary length of the sample led, on the one hand, to an increase in C_{BX} (which in turn causes a sharp drop in the legitimate signal level in signal integration conditions which thereby reduces the possible dynamic range for investigating the photoresponse, see formula (5)) and, on the other hand, causes an increase in absorption in the test crystal, which may cause distortions in the dependencies of the photoconductivity signal on the excitation laser emission intensity recorded in the experiment.

With $l_{обр} \sim 0.1$ æ^{-1} realized in our experiment we could neglect distortions introduced by laser emission energy absorption lengthwide along the sample (at 0.27 μm in undoped alkali-halide crystals, for example, $\text{æ} \sim 10^{-1} - 10^{-2}$ cm^{-1} [57]).

The extraordinary increase in the interelectrode distance which would, it would seem, would produce an increase in the sensitivity of the photoconductivity recording circuit to weak signals (of course with a corresponding increase in the cross-sectional area of the laser beam, see (5) and (6)) requires, however, a corresponding increase in the high voltage applied to the sample in order to conserve the same electric field strength in the sample itself, which rapidly causes breakdown between the electrodes and the sample.

In investigating the photoresponse of the dielectrics in addition to regular photoconductivity a signal caused by the change in the dielectric constant of the crystal resulting from its heating by the laser emission could in principle arise. However even with maximum laser emission intensities used in our case ($\sim 10^8$ W/cm^2) the level of absorbed energy over the pulse duration is less than 0.1 J/cm^3 (with a linear absorption coefficient of ~ 0.1 cm^{-1} for alkali-halide crystals at our wavelengths), which corresponds to heating by 0.1 K and therefore this effect may be ignored [58].

Since in these experiments the laser beam diameter d at the entrance to the sample did not, as a rule, exceed 4 mm, the laser emission did not directly impact the electrodes (see Figure 3). We selected thin platinum wafers 2 x 10 mm^2 for use as the electrodes. Compared to electrodes made from other metals (copper, nickel, silver, etc.) the platinum electrodes had the maximum electron work function (the red photoeffect boundary for platinum is approximately 5.3 eV), which exceeds the quantum energy of the neodymium laser fourth harmonic (approximately 4.7 eV). Therefore platinum is virtually the only electrode material that makes it possible to avoid parasitic signals caused by the photoeffect from electrodes in investigating the photoconductivity of dielectrics at such high emission frequencies.

The absence of parasitic signals from possible irradiation of the electrodes by direct or scattered laser emission was established in a special experiment where a scattering silica plate was inserted into the laser path when the scattered emission impacted the electrodes there was no increase observed in the photoconductivity signal compared to the case of total screening of the electrodes from scattered emission.

Moreover at low and medium laser emission intensities, i.e., at intensities where the irreversible photoconductivity growth mechanism is not yet manifest (which will be discussed below, in Chapter 3 of this study) a test experiment was performed by direct simultaneous illumination of the samples and the electrodes with laser emission. The results from this experiment are given in Figure 4 (a KCl crystal sample was used as the test sample). It is clear that irradiating the platinum electrodes with laser emission at 0.27 μm does not produce a parasitic signal and therefore causes no distortion to the photoconductivity signal. Analogous results were obtained from test experiments using other crystals. On this basis we may conclude that the irradiation of electrodes by the weaker laser emission scattered in the test sample bulk does not cause distortion in the recorded laser photoconductivity signal.

The contribution of surface photoconductivity to the legitimate signal was eliminated by placing the end surfaces of the samples outside the electric field applied to the electrodes and eliminating illumination of the lateral facets of the samples.

The experiments aimed at investigating the laser photoconductivity of alkali-halide crystals were carried out with intense illumination of samples in the absorption band of F-centers by a 100 W halogen incandescent lamp. Without such illumination the photoconductivity signal was 3-5 times greater which clearly may be attributed to the formation of color centers under UV laser action. We emphasize that the nature of the $\sigma(I)$ relation was independent of the existence or absence of illumination.

Moreover it was established in a special experiment at 0.27 and 0.53 μm that intense illumination of samples in the absorption band of the F-centers eliminates the color centers formed by UV laser action over the time between laser pulses (10-15 sec). In this experiment by irradiating the test crystal initially with laser emission at 0.27 μm, F-centers were created in the crystal which was determined by the characteristic tinting of the region of the sample impacted by the laser emission. The concentration of color centers in this case was better than 10^{15} - 10^{16} cm^{-3} if estimated by the light absorption (the absorption cross-section of light by the F-centers was $\sigma_F \sim 10^{-16}$ cm^{-2}, and for an absorption coefficient of $æ \sim 0.1$ cm^{-1} and a characteristic length of the test samples of $\sim$ 2 cm such concentrations are the minimum concentrations established from visual observation).

Then laser emission at 0.53 μm was directed to the same region of the sample. Without illumination of the samples a photoconductivity signal was clearly observed in the first 0.53 μm pulses; naturally this may be attributed to the existence of color centers in the crystals, since without preliminary irradiation by emission at 0.27 μm there was no photoconductivity signal in the test samples at λ = 0.53 μm. The existence of this signal, by the way, makes it possible to estimate low color center concentrations.

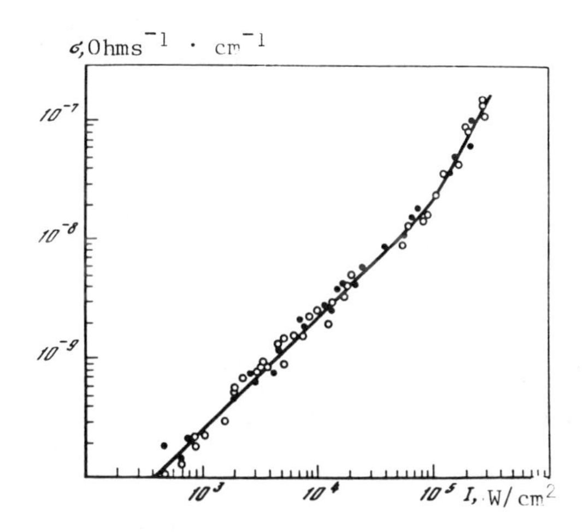

Figure 4. The influence of direct irradiation of electrodes (solid circles) on the photoconductivity signal (using KCl as an example). The open circles represent the photoconductivity without direct illumination of the electrodes; λ = 0.27 μm.

When the test alkali-halide compound crystals were irradiated during the laser pulse by a halogen incandescent lamp in the absorption band of the F-centers the photoconductivity signal at 0.53 μm was not observed even after preliminary irradiation by laser emission at λ = 0.27 μm which indicate the vanishing of color centers.

The maximum error from direct measurements of the values of τ_H, l, S, U; f, C_{BX}, R_H entering into expressions (5) and (6) was 10–20%. The error in determining the photoconductivity signal was less than 10%, i.e., the relative measurement accuracy of the photoconductivity signal was rather high. In determining the directly measured absolute photoconductivity values of the test samples from expressions (5) and (6) we may be off by a factor of several times. The relative error in measuring the intensity of the incident laser emission (determined by the sum of the relative errors in measuring the laser pulse duration, the emission energy and the beam radius at $1/e$ of its maximum along the axis) was less than 40%.

When using broad laser beams (diameter of greater than 2 mm corresponding to the minimum laser emission intensities used in the experiments of $\sim 10^3$ W/cm^2) the capacitor formed by the platinum electrodes and the test dielectric crystal sample can no longer be considered a plane capacitor, since the inhomogeneity of the external electric field at the lateral ends of the electrodes may become manifest. Estimates reveal, however, that in this case the maximum measurement error of the absolute photoconductivity signal did not exceed a factor of 1.5–2 and errored on the high side.

6. Test Crystals

In our experiments we investigated the laser photoconductivity of NaCl, KCl, KBr, KI, RbCl, CsI, CsBr alkali-halide crystals which were obtained from their various sources. Different techniques (the Kyropoulos and Czochralski techniques) were used to grow the crystals. The test samples did not contain any deliberately introduced impurities and were melt grown from OSCh and KhCh brand salts.

We made our selection of the alkali-halide crystals so that their bandgaps (ε_g) were separated to the greatest possible extent. Thus, for example, the most reliable experimentally determined value of ε_g for NaCl crystals was obtained at the boundary of Wannier-Mott excitons and equals $\sim$ 8.8 eV [59] while in CsI $\varepsilon_g \sim$ 6.0–6.4 eV [60].

For our purposes dielectric crystals with a bandgap exceeding two quanta ($2\hbar\omega > 9.4$ eV) in the energy absorption process at the wavelengths used in the experiments were not suitable (i.e., LiF and NaF crystals in which $\varepsilon_g \sim$ 12–14 eV [61–62], since in this case band-to-band carrier excitation by two-photon ionization could not occur even at $\lambda = 0.27$ µm, and possible single-photon ionization of the impurity states was not of independent interest in our case. The probabilities of processes occurring involving two or more photons in a single ionization event ($p \geqslant 3$) are significantly less than when $p = 2$ [51]. Nonetheless in order to test this possibility we investigated the photoconductivity in the UV of CaF_2 crystals having $\varepsilon_g \sim$ 10 eV as well as the photoconductivity of alkali-halide crystals with $\varepsilon_g \sim$ 8–9 eV at 0.35 µm.

In the experiments to investigate laser photoconductivity of alkali-halide crystals we selected a large statistical group for reliable interpretation of the derived data: The number of samples of each test crystal was between 15 and 20, and in certain cases (NaCl, KCl, KBr): As many as several tens of samples.

The entrance and exit crystal surfaces were obtained by shearing along the cleave planes which largely reduced surface absorption.

We should note that in our experiments in the majority of cases we investigated the photoconductivity of alkali-halide crystals whose optical breakdown thresholds we had investigated previously (including in the UV) [20]. The results presented in this chapter were also obtained on samples having a higher optical radiation strength at various wavelengths [7] as well as on samples exposed to reinforcing heat treatments and revealing reproducible temperature and frequency dependencies of the breakdown thresholds [3, 6]. As a result the concentrations of uncontrolled impurities and inclusions in the test samples must be minimized [6, 7] compared to crystals with lower breakdown thresholds.

CHAPTER 3

EXPERIMENTAL RESULTS ON THE LASER PHOTOCONDUCTIVITY OF WIDE-BAND DIELECTRICS

The present chapter presents results from experimental investigations of the laser photoconductivity of wideband dielectrics in the UV (0.27 and 0.35 μm) based on the dependencies of the photoconductivity on the excitation laser emission obtained over a wide range of emission intensities.

Alkali-halide compound monocrystal samples were investigated having various bandgaps ($\varepsilon_g \sim 6$-9 eV) together with KDP, DKDP, ruby and fluorite crystals.

7. Photoconductivity of Alkali-Halide Crystals at 0.27 μm

When the alkali-halide crystals were exposed to the fourth harmonic of a neodymium laser ($\lambda = 0.27$ μm, $\tau_H \sim 8$ nsec) we observed a photoresponse corresponding to an increase in their conductivity. The lifetime of photoconductivity when recording the signal in the high resolution mode did not exceed the laser pulse duration indicating that in our photoexcitation conditions the carrier lifetime was shorter than $8 \cdot 10^{-9}$ sec. This does not contradict results obtained in [37, 38] where investigating defect formation in alkali-halide crystals by picosecond spectroscopy at this same wavelength revealed a nonequilibrium carrier lifetime of $\leqslant 5 \cdot 10^{-11}$ sec (at carrier densities of $n \sim 10^{17}$ cm^{-3}), together with the results given in [25] where under photoexcitation at $\lambda = 0.35$ μm the carrier lifetime was 10^{-9} sec (for $n \sim 10^{10}$ cm^{-3}).

The amplitude of our photoresponse pulse was proportional to the amplitude of the high voltage applied to the samples when the voltage varied over a range of 0.2-12 kV. Such a linear response of the photoconductivity signal to a change in voltage across the electrodes indicates linearity of the photoresponse measurement technique with capacitive coupling between the electrodes and the sample throughout the entire measurement circuit. Our laser photoconductivity measurement technique made it possible to obtain electric signals over a rather broad range: From 100 μV to 10-20 V. The characteristic photocurrent value corresponding to the minimum observed signal was approximately 10^{-7} A.

In our experiments we focused on the derivation and analysis of the dependencies of the photoconductivity signal on the incident emission intensity. Investigating the nature of such relations over a broad intensity range (10^3 - 10^8 W/cm^2) makes it possible to obtain important information on the carrier photoexcitation processes.

Figure 5, a, gives an experimental plot of the photoconductivity (σ) against the excitation laser emission intensity (I) at λ = 0.27 μm obtained for one of the KCl crystal samples. For simpler interpretation of the experimental data here and henceforth the photoconductivity related to the photoconductivity signal by relation (5) is plotted on the Y-axis. The σ(I) relation noted here was recorded over a rather small intensity range (~ 10^3 - 10^5 W/cm^2). It is clear from Figure 5, a, that in this range of intensities $\sigma \sim I$.

Analogous linear or near-linear σ(I) relations were obtained at this wavelength in the intensity range noted for other KCl samples as well and for other types of alkali-halide crystals. Figure 5, b, illustrates this using a sample NaCl crystal while Figure 5, c, shows a sample KBr crystal. The absolute photoconductivity signal level at λ = 0.27 μm in the various alkali-halide crystals and different samples of the same alkali-halide crystal varied over a single order of magnitude.

Such a nature of the σ(I) relation in the low intensity range may evidently be associated with the single-photon ionization process of the impurity or defect states.

When the alkali-halide crystals are irradiated by medium intensity laser emission (~ 10^5 - 5 · 10^6 W/cm^2) at λ = 0.27 μm the nature of the σ(I) relation varied significantly: The relation became nonlinear and near-quadratic. The observed slope of the curves running through the experimental points of the σ(I) relation characterizing on our double logarithmic scale the power in intensity may vary over a certain range (~ 1.7-2.1) when investigating both various samples of the same crystal and different alkali-halide crystals. This is due to the relative narrowness of the range of emission intensities in which nonlinear σ(I) relations have been observed in alkali-halide crystals and the existence of transition regions from one section of the relation to another.

Figure 6 gives σ(I) relations measured in the medium emission intensity range at λ = 0.27 μm for KCl crystal samples. Figure 7 (in the same range of intensities) presents $\sigma(I)$ experimental relations obtained for one of the NaCl crystal samples (a) and for a KBr crystal sample (b). $\sigma(I)$ relations had an identical nature in the medium intensity range ($\lambda = 0.27$ μm) for our other alkali-halide crystals as well. The absolute photoconductivity level at a certain fixed intensity in this case varied insignificantly, within half an order of magnitude.

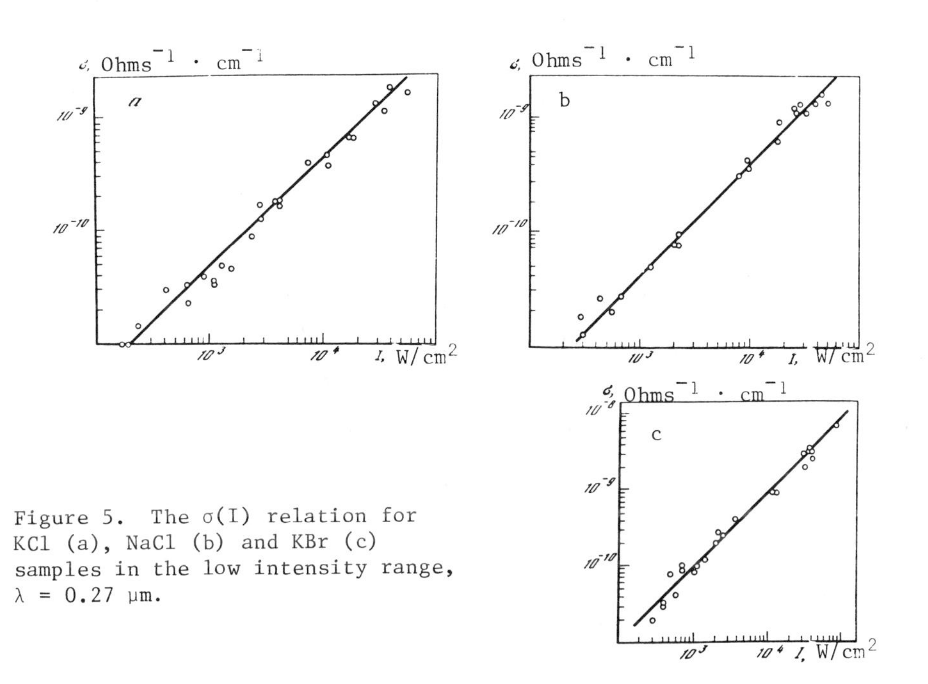

Figure 5. The $\sigma(I)$ relation for KCl (a), NaCl (b) and KBr (c) samples in the low intensity range, $\lambda = 0.27$ μm.

Here we consider it necessary and important to note the following fact for our further discussion. The $\sigma(I)$ relations given for the medium emission intensity range in Figure 7 were obtained not only as I varied from low values to high values (in a "forward" progression), but also, in one experiment, with a reverse change in I: From high values to low values. Here as indicated by the data the nature of the $\sigma(I)$ relations does not change and the experimental points of both the forward and reverse progression lie on a virtual identical curve.

We observed an analogous picture for other alkali-halide crystals as well and in experiments conducted in a low intensity range, only if the crystal samples were not previously irradiated by a more powerful UV source than in the two regions noted above.

In interpreting the results from experiments on laser photoconductivity of alkali-halide crystals in the medium emission intensity range at $\lambda =$ $= 0.27$ μm we may state that the nature of the $\sigma(I)$ relation we observed (near-quadratic) is either the result of the cascade excitation of carriers in the transition from the valence band of the crystal to the conduction band through intermediate impurity state levels or as a result of two-photon band-to-band ionization. We note that the authors of [63-65] were among the first to observe two-photon ionization in undoped alkali-halide crystals. In recent years the investigation of two-photon lattice absorption has attracted the attention of an increasing number of researchers, since it may provide much new information on the band structure of solids [66-68].

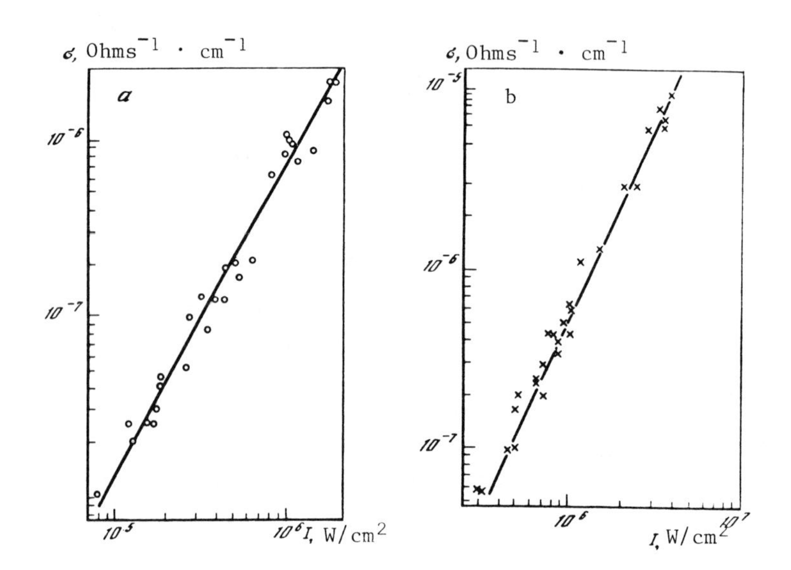

Figure 6. The $\sigma(I)$ relation for KCl samples in the medium intensity range at $\lambda = 0.27$ μm.

a: $\sigma \sim I^{1.7}$; b - $\sigma \sim I^{2.1}$.

At higher emission intensities ($\sim 10^7 - 10^8$ W/cm^2) the nature of the dependence of the photoconductivity signal again changed: The $\sigma(I)$ relation became linear and the change itself occurred in a rather narrow range of intensities: $I \sim (2\text{-}6) \cdot 10^6$ W/cm^2. Figure 8 gives sample experimental $\sigma(I)$ relations measured in the medium and high intensity ranges (at $\lambda =$ $= 0.27$ μm) for NaCl and CsBr crystal samples. The $\sigma(I)$ relation remained linear up through the surface breakdown threshold of the test crystals.

In this range of high emission intensities these data on the laser photoconductivity may be attributed either to the cascade transition of the carriers through the intermediate impurity levels lying in the bandgap from the saturation of one of the transitions, or by the significant influence of quadratic recombination processes (with a nonequilibrium carrier concentration $n \geqslant 10^{12}\ cm^{-3}$) in which the recombination time is $\tau_{рек} \sim n^{-1}$.

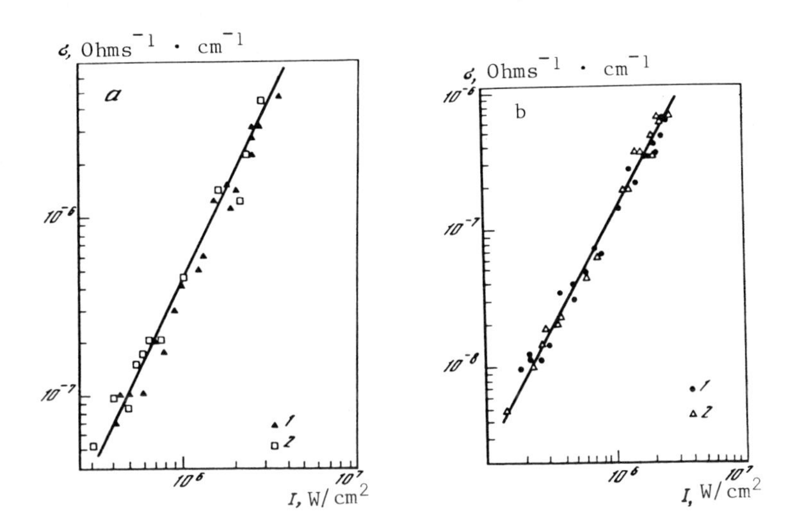

Figure 7. The $\sigma(I)$ relation for NaCl (a) and KBr (b) samples in the medium intensity range for forward (1) and reverse (2) events; λ = 0.27 μm.

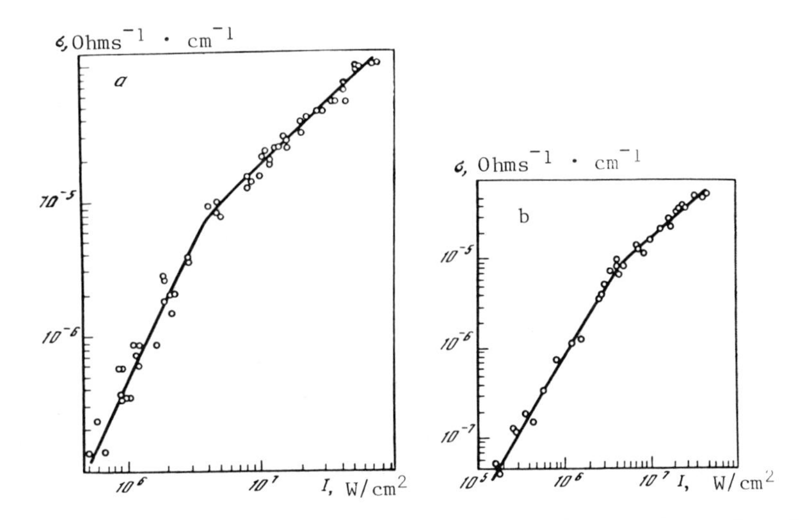

Figure 8. The $\sigma(I)$ relation in the transition from medium emission intensities to high emission intensities in NaCl (a) and CsBr (b); λ = 0.27 μm.

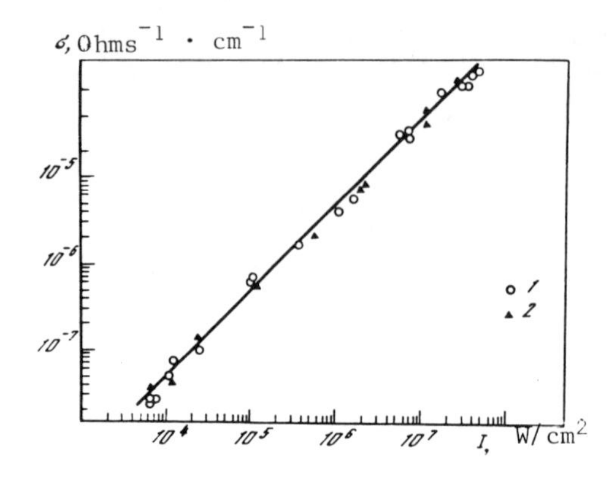

Figure 9. The photoconductivity of an RbCl sample plotted as a function of excitation emission intensity at $\lambda = 0.27$ μm after the sample has been irradiated at the same wavelength but with $I \geqslant 10^7$ W/cm^2 for the forward (1) and reverse (2) events.

If alkali-halide crystals initially exposed to ultraviolet radiation at $I \geqslant 10^7$ W/cm^2 are then exposed to laser emission at $\lambda = 0.27$ μm and the σ(I) relations are recorded so that the change in intensity occurs from large values to small values, nonreproducibility appears in the initial σ(I) relation which manifests a hysteresis nature and the nonlinear section characteristic of medium intensities in this relation vaniches and σ(I) becomes very near linear up through the region of low values of I.

The photoconductivity signal levels obtained in the reverse (in intensity) progression continue to rise with respect to the signals first obtained in the forward progression, while the σ(I) relation itself is a continuation of its linear section corresponding to the high intensity range. This observed growth of photoconductivity in alkali-halide crystals is stable in time and is irreversible, i.e., it is conserved for at least many tens of hours even under long-term irradiation of the crystals by a powerful silica lamp emitting over a broad frequency range. When the test sample is again irradiated by laser emission of any intensity in the test range (at 0.27 μm) σ(I) does not change and now remains linear for both the forward and reverse intensity progressions. Figure 9 illustrates the nature of the σ(I) relation after irradiation by powerful UV emission at $\lambda = 0.27$ μm using a sample RbCl crystal.

The relative magnitude of the irreversible rise of photoconductivity for different alkali-halide crystals and for samples of the same crystal was most significant in the comparatively low intensity range of laser emission (~ 10^3 - 10^4 W/cm^2) and varied over a factor of 10-40.

It is important to emphasize the universality of this effect with respect to all alkali-halide crystals we investigated at 0.27 μm. Moreover the separate characteristic details of the σ(I) relation observed in the experiment could differ among themselves. Thus, for example, Figure 10 gives sections of this relation for KCl and NaCl crystal samples.

It is clear from a comparison of these graphs that in the case of NaCl in the progression from low intensities to high intensities the transition from a nonlinear σ(I) relation to a linear relation occurs at somewhat higher emission intensities, while the registered onset of the irreversible increase in photoconductivity from the reverse progression (indicated by the dotted line and the arrow in the diagrams) corresponds to $I \sim 6 \cdot 10^5$ W/cm^2 in the NaCl sample and $I \sim 2 \cdot 10^6$ W/cm^2 in the KCl sample. Therefore a very narrow transmission range is observed from one section of the $\sigma(I)$ relation to another. A typical intensity value at which this transmission occurs for many alkali-halide crystals is $I \sim 5 \cdot 10^6$ W/cm^2.

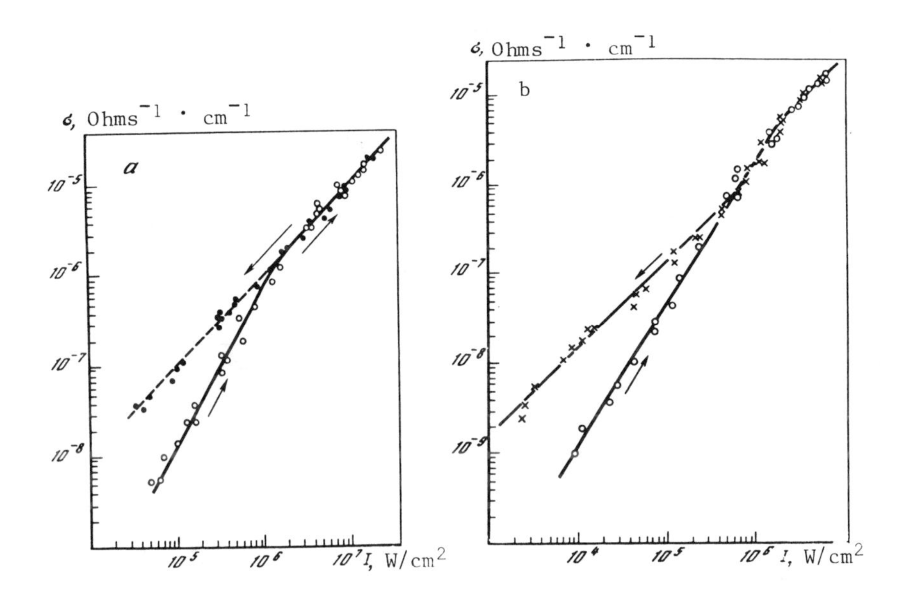

Figure 10. The irreversible increase in photoconductivity in KCl (a) and NaCl (b) samples after exposure to high intensity radiation.

The forward and reverse progressions are indicated by the arrows; $\lambda = 0.27$ μm.

The critical position of the linear $\sigma(I)$ relation corresponding to the greatest irreversible increase in photoconductivity in the transition from high intensities to low intensities is entirely determined by the slope of the $\sigma(I)$ relation on the second linear section (corresponding to values of $I \geqslant 10^7$ W/cm^2). In addition we note that in the reverse progression we never observed a sublinear intensity dependence of the photoconductivity. We may take the unique optical hysteresis effect discovered in our experiments and the related features in the $\sigma(I)$ relation to be attributable to generation of stable radiation defects in the test alkali-halide crystals under high-power UV laser radiation.

We performed the following experiment in order to establish the possible influence of accumulation effects on the nature of the σ(I) relation. In the low and medium intensity ranges (at $\lambda = 0.27$ μm) where the irreversible increase in photoconductivity is not yet clearly manifest, the alkali-halide crystal samples were exposed to multiple laser irradiation (several hundred laser pulses for each point). The dependence of the photoconductivity signal on intensity was again measured after this process. The results from this experiment for one sample KCl crystal are shown in Figure 11. It is clear that there is no "signal accumulation" with such multiple irradiation in either the low or medium intensity ranges.

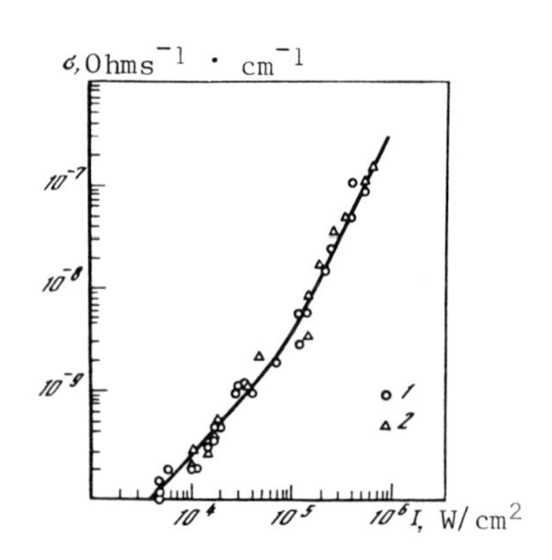

Figure 11. Photoconductivity of KCl at $\lambda = 0.27$ μm before (1) and after (2) preliminary irradiation of the sample by a series of M pulses.

$M \geqslant 100$, $I \sim 10^4 - 5 \cdot 10^5$ W/cm^2.

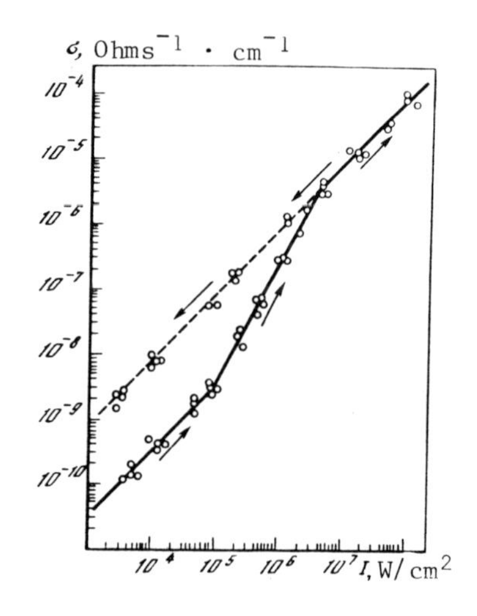

Figure 12. The σ(I) relation for KCl obtained over a broad range of intensities in a single experiment.

The arrows indicate the direction of change in intensity; $\lambda = 0.27$ μm.

This fact indicates that by multiple irradiation of samples with radiation of relatively low intensity ($I < 10^6$ W/cm^2) it is not possible to achieve an irreversible increase in their photoconductivity, which confirms the hypothesis above of defects arising in the test crystals only from irradiation by high-intensity laser emission.

The results from research on the photoexcitation of nonequilibrium carriers based on an investigation of the nature of the variation in the σ(I) relation at $\lambda = 0.27$ μm carried out in a single experiment for the same alkali-halide crystal sample contain the same features as discussed above for various experiments conducted using different samples.

Figure 12 shows a sample $\sigma(I)$ relation for a KCl crystal that is typical of all test alkali-halide crystals at 0.27 μm obtained in one experiment. Three characteristic sections may be observed in this relation in the progression from low intensities to high intensities: A linear section at $I \sim 10^3 - 5 \cdot 10^4$ W/cm^2, a near-quadratic section at $I \sim 10^5 - 5 \cdot 10^6$ W/cm^2 and a second linear section at $I \sim 5 \cdot 10^6 - 10^8$ W/cm^2. It is clear from Figure 12 that in the reverse progression from high radiation intensities to low intensities, the $\sigma(I)$ relation had a hysteresis nature while the photoconductivity signal was not as it appeared in the quadratic section of the $\sigma(I)$ relation, but rather exceeded this value. The relative excess value measured at $I \sim 5 \cdot 10^3$ W/cm^2 was as high as a factor of 25 in this case. Subsequent irradiation of the sample by laser emission at $\lambda = 0.27$ μm did not change the nature of the $\sigma(I)$ relation: This relation remained linear both for the forward and reverse progressions.

It should be noted that the absolute values of the photoconductivity of the test crystals determined by expressions (5) and (6) make it possible to carry out a quantitative estimate of such an important parameter to wideband dielectrics as the nonequilibrium carrier concentration. The importance and critical nature of the determination of n was noted, specifically, in our studies [41] and also in [55] which investigated the critical optical laser radiation strength of matter based on the laser photoconductivity in the UV.

If we assume an electron mobility at room temperature for alkali-halide crystals of 10 cm$^2 \cdot$ V^{-1} sec^{-1} [69] with a photoconductivity of the test samples of the order 10^{-6} Ohms$^{-1} \cdot$ cm^{-1} we find $n \sim 10^{12}$ cm^{-3}. Since the nonequilibrium carrier concentration in this case is determined with mobility accuracy, whose absolute measurements in the dielectrics have significant discrepancies (see, for example, [1, 60]), the accuracy in determining n is one to two orders of magnitude.

8. The Photoconductivity of Alkali-Halide Crystals at 0.35 μm

By irradiating alkali-halide crystal samples with laser emission at the third harmonic of a neodymium laser (pulse duration $\tau_И \sim 8$ nsec) we also observed a photoresponse whose polarity indicated an increase in the conductivity of the samples: A positive photoresponse. The duration of the photoresponse did not exceed the laser pulse duration which is in agreement with the data provided in [25] where at the same frequency at lower carrier densities ($n \sim 10^{10}$ cm^{-3}) and at $\tau_И \sim 20 \cdot 10^{-12}$ sec the nonequilibrium carrier lifetimes were measured in KBr and NaF crystals at $\sim 10^{-9}$ sec.

In our experimental conditions the laser photoconductivity of KCl, NaCl and RbCl crystals at $\lambda = 0.35$ μm even at maximum emission intensities in the experiment were three to four orders of magnitude smaller than at $\lambda = 0.27$ μm. These crystals are dielectrics with the broadest bands of all our test samples. For example in KCl the bandgap is reliably determined in a number of experimental studies [70, 71]: $\varepsilon_g \sim 8.7$ eV. In an RbCl crystal $\varepsilon_g \sim 8.3$ eV [72] and the authors have noted that in this case the experimental data are not in agreement with calculation.

The noticeable photoconductivity at $\lambda = 0.35$ μm making it possible to investigate its behavior as a function of the change in intensity of the excitation emission over a rather broad range of values appeared in these crystals only after preliminary irradiation at $\lambda = 0.27$ μm. Preliminary irradiation involved single-event or multiple (5-10 laser pulses) irradiation of test samples by emission with an intensity of $I \geqslant 10^7$ W/cm^2. The intensity dependence of photoconductivity at $\lambda = 0.35$ μm in this case is linear across the entire test range of intensities. The absolute value of the photoconductivity signal in this case is near that of the same alkali-halide crystal at $\lambda = 0.27$ μm after preliminary bombardment of the samples with high intensity radiation. We emphasize that the $\sigma(I)$ relation at $\lambda = 0.35$ μm for alkali-halide crystals with $\varepsilon_g \sim 8.3$-8.8 eV conserves its linear nature and is reproducible for both the forward and the reverse change in intensity. Figure 13 demonstrates this process using one NaCl crystal sample.

Multiple (up to 100 laser pulses) irradiations of the crystal by low intensity radiation at $\lambda = 0.35$ μm will not change the photoconductivity signal value.

If alkali-halide crystals having $\varepsilon_g \sim 8.3$-8.8 eV are first irradiated by emission of intensity $I \ll 10^7$ W/cm^2 at $\lambda = 0.27$ μm, the photoconductivity signal level in these crystals at $\lambda = 0.35$ μm remains the same as observed in samples that are not exposed to preliminary irradiation, which does not allow investigation of the $\sigma(I)$ relation over a broad intensity range.

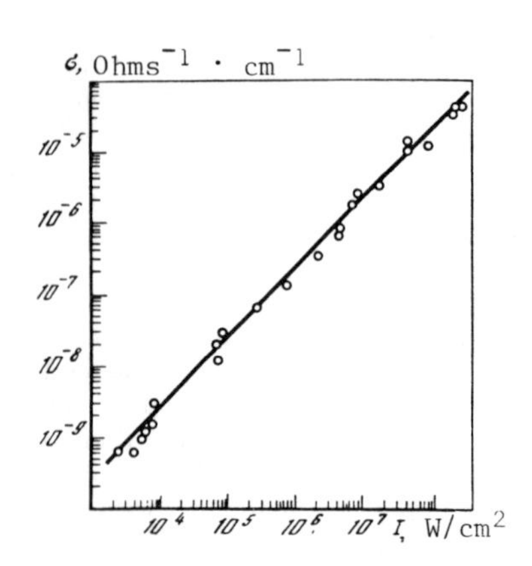

Figure 13. The photoconductivity of an NaCl crystal irradiated by emission at $\lambda = 0.35$ μm after preliminary irradiation at $\lambda =$ $= 0.27$ μm.

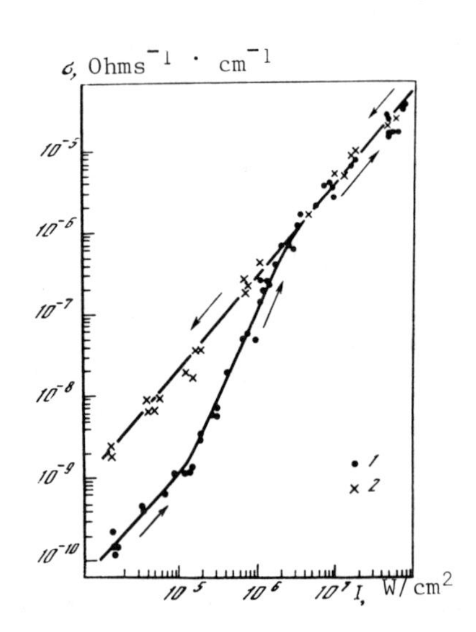

Figure 14. The photoconductivity of KI at $\lambda = 0.35$ μm.

The forward (1) and reverse (2) intensity progressions are indicated by the arrows.

The results on the laser photoconductivity of alkali-halide crystals with a bandgap of 8.3-8.8 eV at 0.35 μm provided here are in principle in good agreement with the assumption in the preceding section that under irradiation by high-power emission with a quantum energy $\hbar\omega \sim 4.7$ eV ($\lambda = 0.27$ μm) energy states arise in the test samples resulting from the formation of radiation defects whose occupation and emptying kinetics produce the linear $\sigma(I)$ relation observed in experiment across the entire range of intensities. Clearly the quantum energies at $\lambda = 0.35$ μm ($\hbar\omega \sim 3.5$ eV) are not sufficient to cover the bandgap of these wideband dielectrics in the two-quantum energy absorption process. Since radiation defect formation occurs through the electron states in the conduction band (see [37, 38]), under irradiation at $\lambda = 0.35$ μm virtually no defects are formed in the crystals discussed above.

A previously cited study [25] was devoted to an investigation of photoconductivity in KBr at 0.35 μm. This study derived the dependence of the charge induced across the electrodes on the laser emission intensity producing photoelectrons in a picosecond pulse range. However when using such narrow pulses ($\tau_И \sim 20$ psec) even when employing a signal preamplifier it is not possible to investigate this relation with intensity variations exceeding two to three orders of magnitude, and the relatively low intensity range becomes inaccessible ($\sim 10^3 - 10^5$ W/cm^2).

We investigated [41] the photoconductivity of KI, KBr, CsI and CsBr crystals at $\lambda = 0.35$ μm over a broad range of laser emission intensities ($10^3 - 10^8$ W/cm^2). The bandgap reliably determined in experiment amounts to 6.3 eV in KI [63, 73] at the same time that the experimental values for KBr of $\varepsilon_g \leqslant 7.3$ eV [70] are less reliable and are not in good agreement with calculation results from [74] which yield significantly lower values of $\varepsilon_g \sim 6.0$ eV.

In these crystals with $\varepsilon_g \sim 6$-7 eV the nature of the dependence of photoconductivity on the excitation laser emission intensity at $\lambda = 0.35$ μm remain the same as in the case of $\lambda = 0.27$ μm, i.e., the $\sigma(I)$ relation in the forward progression consisted of three sections: $\sigma \sim I$ in the $I \sim 10^4 - 10^5$ W/cm^2, $\sigma \sim I^2$ at $I \sim 10^5 - 10^6$ W/cm^2 and, finally, $\sigma \sim I$ for $I > 10^7$ W/cm^2.

In the reverse progression of the $\sigma(I)$ relation (as for the case where $\lambda = 0.27$ μm) an irreversible increase in the photoconductivity signal was observed. Figure 14 illustrates this process for a KI crystal. The absolute value of the photoconductivity signal was close to the value we recorded for the same crystals at $\lambda = 0.27$ μm.

These experimental data on the laser photoconductivity of relatively more narrowband alkali-halide crystals (with $\varepsilon_g \sim 6$-7 eV) at $\lambda = 0.35$ μm also confirm the previous assumption that high-power UV laser radiation creates radiation-resistant defects in the test samples which in turn results in an irreversible growth of photoconductivity. The observed nature of the $\sigma(I)$ relation in the forward intensity progression clearly reveals initially single-photon carrier photoexcitation processes (for $I \leqslant 10^5$ W/cm^2) and subsequently (at $I \sim 10^5 - 10^7$ W/cm^2) two-photon carrier photoexcitation processes. A linear $\sigma(I)$ relation in the high intensity range ($I \geqslant 10^7$ W/cm^2) may be attributed to quadratic carrier recombination.

9. The Photoconductivity of KDP and DKDP Crystals

KDP and DKDP crystals have been used successfully in various nonlinear optical devices and systems as laser frequency multipliers; these crystals withstand significant power densities (to 10^9 W/cm^2) and have a rather high emission conversion factor (up to 50%) [75]. A recent study [76] has reported a conversion factor in KDP of approximately 90%.

The bandgap of KDP and DKDP is ~ 9 eV [77], while the dielectric relaxation time at room temperature is approximately 0.1 sec [56]. Therefore $2\hbar\omega > \varepsilon_g$ for KDP and DKDP crystals ($\hbar\omega \sim 4.7$ eV is the quantum energy at $\lambda = 0.27$ μm), while the space charge resulting from photoexcitation relaxes rapidly.

KDP and DKDP crystals 5x5x20 mm^3 in size were cleaved to eliminate the nonlinear optical rectification effect [78]. It follows from the data given in [79] that the linear absorption coefficient at $\lambda = 0.27$ μm for KDP crystals is $æ_l \sim 0.05$-0.1 cm^{-1}. Transmission measurements of the test KDP and DKDP crystals at this wavelength carried out on a spectrophotometer yield similar values of ~ 10^{-2} cm^{-1} which indicates insignificant lattice absorption in KDP and DKDP. With such low absorption coefficients the various effects associated with crystal heating by radiation such as the induced pyroelectric effect [78] which may in principle extend the photoresponse, are virtually absent.

In irradiating the KDP and DKDP crystals with laser emission at $\lambda =$ $= 0.27$ μm a photoresponse was observed whose polarity, as in the alkali-halide crystals, corresponded to an increase in the conductivity of the test samples (a positive photoresponse). The photocurrent pulse duration was 30 nsec, i.e., a value somewhat greater than τ_H. We note that our measurements are in agreement with the data presented in [39] where in similar conditions at $\lambda = 0.53$ μm an upper nonequilibrium carrier lifetime limit of $\leqslant 3 \cdot 10^{-8}$ sec was determined for KDP (DKDP).

The amplitude of the photoresponse was proportional to the amplitude of the external high-voltage applied to the electrodes. In investigating the photoconductivity of KDP and DKDP we were primarily interested, as in the case of the alkali-halide crystals, in the nonequilibrium carrier kinetics over as broad a range of emission intensities as possible.

Figure 15 presents our experimental dependencies of photoconductivity on the laser radiation intensity at $\lambda = 0.27$ μm for the KDP and DKDP crystals. These relations clearly reveal two sections that are obviously related to the different physical processes of carrier photoexcitation: A linear section in the relatively low intensity range (~ 10^3 - 10^6 W/cm^2) and a subsequent quadratic region in the higher intensity range (~ 10^6 - - 10^8 W/cm^2). The direction of the change in laser emission intensity (from low intensity to high intensity or vice versa), unlike the case of alkali-halide crystals, had no influence on the σ(I) relation or on the level of the observed photoconductivity signal. The absolute value of the photoconductivity signal in KDP (DKDP) was approximately one or one and a half orders of magnitude smaller than in the alkali-halide crystals.

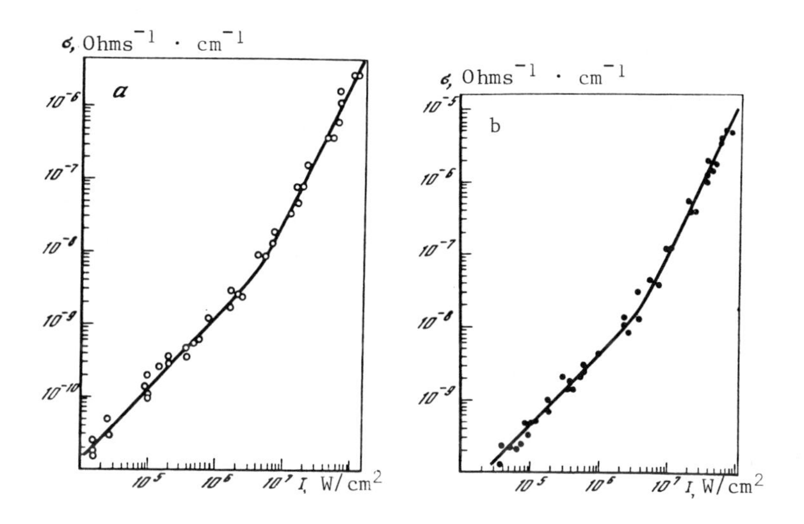

Figure 15. The σ(I) relation for KDP (a) and DKDP (b); $\lambda = 0.27$ μm.

The linear section of the σ(I) relation for the KDP and DKDP crystals at $\lambda = 0.27$ μm may be attributed to single-photon ionization of the impurities or defects. The square-law region of the σ(I) relation may be interpreted as the manifestation of two-photon energy absorption processes.

We may estimate carrier mobility based on data on the photoconductivity of KDP (DKDP). We may represent the nonequilibrium carrier concentration as

$$n = \sigma m^*/(e^2 \tau_{\text{рел}}), \tag{7}$$

where m^* is the electron effective mass, e is electron charge, $\tau_{\text{рел}}$ is the time between electron-phonon collisions.

The quantity $2n\hbar\omega$ represents the number of quanta absorbed in the two-photon ionization of the lattice atoms at the test frequency. On the other hand subject to expression (1) the energy absorbed per unit of volume of the test sample over the pulse duration is $æ_2 I^2 \tau_{\text{и}}$. Therefore bearing in mind (7) and (5) the two-photon absorption coefficient is

$$\text{æ}_2 = 2\hbar\omega m^* u C_{\text{вх}} f^2 / (I^2 \tau_{\text{и}} e^2 \tau_{\text{рел}} U \tau_1 S l). \qquad (8)$$

Here τ_1 is the lifetime of photoconductivity, i.e., the greater time of $\tau_{\text{и}}$ and $\tau_{\text{рек}}$ (recombination time).

If we use the value $\text{æ}_2 = 2.7 \cdot 10^{-4}$ cm/MW for KDP, experimentally determined in [49] by direct measurement based on light transmission and if we set $\tau_{\text{рек}} > \tau_{\text{и}}$ (our observed photoresponse duration of ~ 30 nsec in KDP supports this assumption), then the estimate from expression (8) carried out at intensities corresponding to the square-law section in the $\sigma(I)$ relation yields a carrier mobility value ($\mu = e\tau_{\text{рел}}/m^*$) of the order 10^{-2} $cm^2 \cdot V^{-1} \cdot sec^{-1}$.

We have not encountered data on carrier mobility in KDP (DKDP) in the literature, although our derived value is very close to the measured mobility for sapphire: $\mu \sim 5.2 \cdot 10^{-2}$ $cm^2 \cdot V^{-1} \cdot sec^{-1}$ [80]. On the other hand it is clear that if we have the quantity corresponding to carrier mobility in KDP, we may estimate the two-photon absorption coefficient from expression (8).

10. The Photoconductivity of Ruby and Fluorite Crystals

We detected photoconductivity in fluorite crystals at $\lambda = 0.27$ μm only at very high emission intensities (~ 10^8 W/cm^2). The range of excitation emission intensities used to generate photoconductivity was therefore quite narrow, since there was an upper limit on the surface radiation strength of the fluorite samples. Bearing in mind that the bandgap in CaF_2 ~ 10 eV, our results may be attributed to the weakly-expressed impurity photoconductivity of the test samples. Moreover the derived data indicate that we are not observing processes that involve the absorption of three or more photons in a single ionization event.

In ruby crystals the $\sigma(I)$ relation was linear across the entire range of laser emission intensities (at $\lambda = 0.27$ μm) for either direction of change in intensity which clearly is related to the single-photon ionization of Cr^{3+} impurity levels.

Moreover with multiple irradiation of the ruby samples by UV radiation of constant intensity at $\lambda = 0.27$ μm a reduction in photoconductivity was observed. When the external electric field applied to the sample was switched off, the photoconductivity signal changed polarity and subsequently with growth in the number of laser pulses, the signal vanished (Figure 17). Bearing in mind the relatively low ionic conductivity of ruby at room temperature (~ 10^{-16} $Ohms^{-1} \cdot cm^{-1}$ [81]), and the resulting significant dielectric relaxation time (~ 10^3 sec), the observed effect may be attributed to the formation of a space charge in the ruby whose field strength is opposite the vector of the external field strength.

We note that this effect makes it possible to experimentally determine the specific conductivity of ruby by measuring the rate of space charge relaxation. When the external field is switched off this is achieved by occasional probing of the sample with light pulses so that the space charge relaxation is determined by the intrinsic bulk conductivity of the crystal rather than by the photoconductivity. We also emphasized that by ignoring this effect we may significantly distort the σ(I) relations obtained for crystals with a low (like ruby) bulk conductivity. The proposed method of determining the intrinsic conductivity of dielectric crystals with a rather high Maxwellian relaxation time ($\geqslant 10^2 - 10^3$ sec) is free of many of the drawbacks inherent in other techniques and related either to the use of non-ohmic contacts or to an undesirable contribution from surface conductivity to the measured signal.

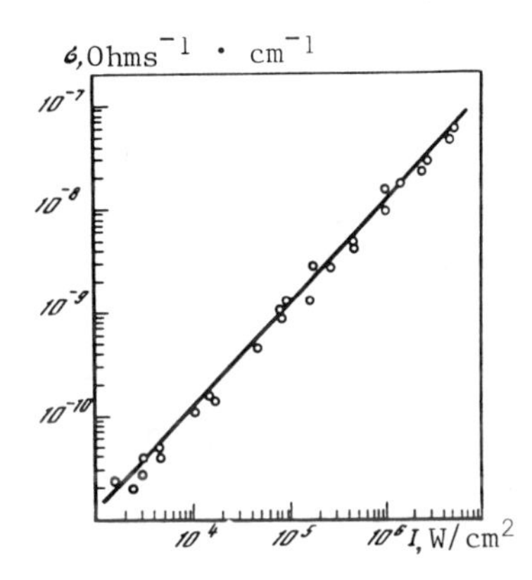

Figure 16. The σ(I) relation for ruby; λ = 0.27 μm.

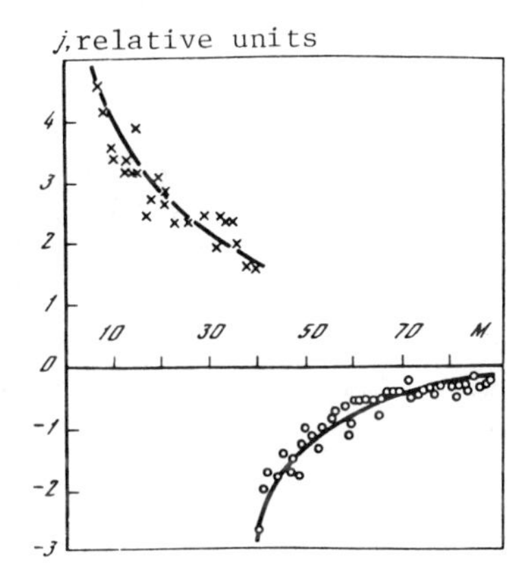

Figure 17. The photocurrent in ruby under irradiation at λ = 0.27 μm as a function of the number of laser pulses (M) of constant intensity ($I \sim 10^7$ W/cm^2). When M ⩾ 40 the external field was switched off.

CHAPTER 4

INVESTIGATION OF THE CARRIER PHOTODRAG EFFECT AND NONLINEAR EMISSION ABSORPTION

The present chapter presents results from experimental investigations of the carrier photodrag effect discovered in alkali-halide crystals under irradiation by high-power UV laser emission (λ = 0.27 μm). This chapter also covers experimental studies of nonlinear absorption of laser radiation energy in alkali-halide crystals over a broad range of intensities (λ = = 0.35 and 0.27 μm) based on optoacoustic techniques.

Results from such investigations in connection with photoconductivity data for identical alkali-halide crystals at the same wavelengths made it possible to more reliably interpret the nonequilibrium carrier photo-excitation mechanisms.

11. The Photon Drag Effect on Electrons in Alkali-Halide Crystals

When phonons interact with electrons in addition to energy the electrons also acquire the momentum of the photons, which produces ordered current carrier motion with respect to the lattice in the direction of light propagation. The magnitude of the resulting flux will be highly dependent both on the specific energy band structure and the light-carrier interaction mechanism.

The appearance of an electron flux resulting from the electron transfer of momentum from directional photon flow is observed in the optical and microwave ranges in semiconductors, semimetals and other metals. A similar photon drag effect on electrons has been investigated in semiconductors (Ge, Si, $A^{III}B^{V}$-type compounds, etc.).

The momentum transfer process from the photons to the free carriers may occur with simultaneous participation of a third particle in addition to the electron and the photon (such as another photon or the impurity center [82]). In the case of such indirect light absorption the photon momentum in the interaction process itself is partially conveyed to the lattice. Another possibility is that only two particles participate in the initial interaction process: The photon and the electron. Such a process occurs in the production of recoil electrons from Compton scattering. In solids such interaction between two particles will also occur from direct transitions between energy bands.

An experimental study of the drag effect became possible only with the development of lasers. It was possible to observe several different varieties of the effect in doped semiconductors (primarily in germanium): Drag on electrons from intraband absorption of light by the free carriers [83]; in band-to-band transitions [84], and the photoionization of impurities [85].

Investigating the characteristics of the drag effect for application to dielectric crystals may provide very valuable information on nonequilibrium carriers, particularly the magnitude of such an important and elusive parameter for dielectrics as the electron recombination time. In this regard we undertook experimental investigations of the drag effect on nonequilibrium carriers arising in wideband dielectrics under irradiation by high-power UV laser emission [86]. We employed the single-mode laser described in Chapter 2 in the experiments.

We employed undoped alkali-halide crystal (NaCl, KCl, KBr) samples 4x4x20 mm^3 grown from a melt of brand OSCh salts as the test samples. The emission power density was 10^8 W/cm^2 with a weakly-focused beam diameter of 1 mm (at a level of $1/e$ of the axial intensity), which is below the breakdown threshold of the crystal surfaces [53]. The beam diameter was measured using the method proposed in [87, 88].

Platinum electrodes were placed on the surface of the samples 2 mm from their ends. We remember that the red boundary of the photoeffect in this metal lies further into the UV than the quantum value in our emission at 0.27 μm. The electrode width was 1 mm with an interelectrode distance of 12 mm. A diaphragm 1 mm in diameter was placed in front of the samples.

A number of experiments were performed to investigate the drag effect. In the first experiment the electrodes were connected to the inputs to two independent high-sensitivity charge amplifiers based on GT-341A transitors with a low input capacitance (of the order $5 \cdot 10^{-12}$ F). The amplifiers were placed as close as possible to the test sample (in actuality at a distance of ~ 50 mm), in order to minimize the input capacitance of the circuit and at the same time to improve its sensitivity. The recording circuit operated in a signal integration mode.

The signals taken from the amplifier outputs were injected to a S8-2 dual-trace storage oscilloscope. All circuit components provided a bandwidth of greater than 7 MHz. Figure 18 gives the organizational configuration of this experiment and a typical oscillogram of the recorded signals.

The experiment revealed that a positive charge arises in the first electrode on the beam path during the laser pulse. A negative charge is generated across the second electrode in this case. The amplitudes of these two opposite signals were virtually identical. When irradiating the test samples on the opposite side the observed current pulses changed polarity. If the laser emission impacted the sample between the electrodes perpendicular to the direction shown in Figure 18, no signals appeared. In this experimental geometry, however, rotating the sample in the plane of the diagram by a certain angle with respect to the direction of laser emission propagation would generate signals having a polarity and amplitude determined by the direction and angle of rotation, respectively.

In the second experiment a third round platinum electrode connected to the amplifier input and analogous to the first two electrodes was connected halfway between the two far electrodes on the samples. The output signal from the third amplifier was injected to a second S8-2 oscilloscope whose second channel was designed to record the laser UV emission energy. The experimental configuration and the oscillograms of the test signals are shown in Figure 19. When the signals of different polarity noted above were observed in the first and last electrodes on the laser beam path, no charge accumulated at the center electrode.

The amplitude and polarity of the signals recorded in these experiments and the conservation of both signal amplitude and polarity with a 180° change in sample orientation with respect to the direction of emission propagation reveals the photon lag effect on the electrons. In this case the direction of electron flow coincides with the direction of propagation of the electromagnetic wave.

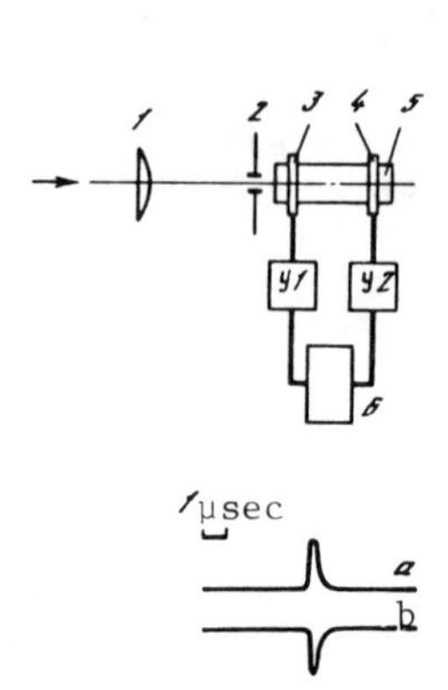

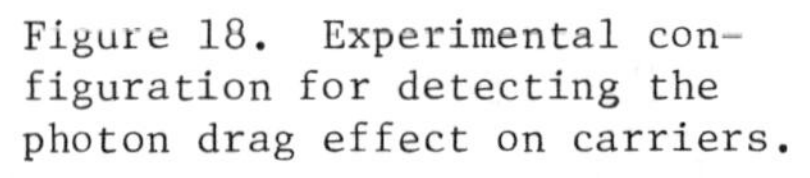

Figure 18. Experimental configuration for detecting the photon drag effect on carriers.

1 - Silica lens, $f_0 \sim 500$ mm; 2 - diaphragm, $d \sim 1$ mm; 3, 4 - circular platinum electrodes; 5 - test sample; Y1, Y2 - electric signal amplifiers; 6 - oscillograph. The arrow indicates the direction of laser emission propagation; $\lambda = 0.27$ μm.

Oscillogram: a - Signal waveform from Y1; b - signal waveform from Y2.

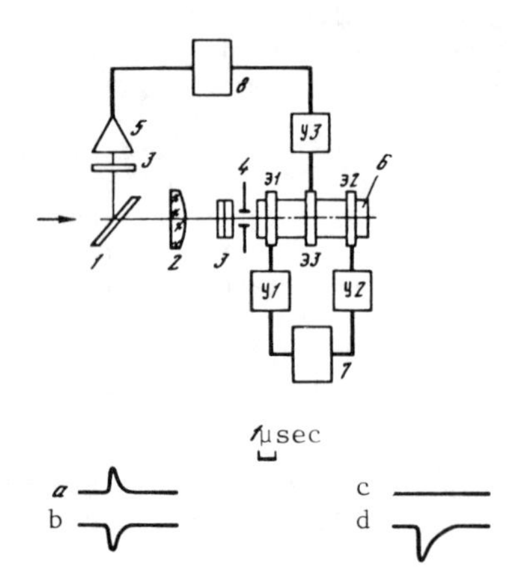

Figure 19. Experimental configuration for investigating the drag effect.

1 - Silica plate; 2 - silica lens, $f_0 \sim 500$ mm; 3 - light filters; 4 - diaphragm; 5 - pyrodetector for energy monitoring; 6 - sample; 7, 8 - oscilloscopes; Э1, Э2, Э3 - platinum electrodes; Y1, Y2, Y3 - signal amplifiers. The arrow indicates the direction of emission propagation; $\lambda = 0.27$ μm.

Oscillogram: a - Signal waveform from Y1; b - signal waveform from Y2; c - zero signal from Y3; d - laser pulse (integrated).

In the last experiment one of the two end electrodes was grounded, while the other was connected to the second amplifier whose output signal was injected to an S8-2 oscilloscope. The second oscilloscope channel was designed to record the energy of the UV laser emission. The dependencies of the charge (q) induced across the ungrounded electrode on the laser emission intensity at $\lambda = 0.27$ μm were recorded. Figure 20 gives the q(I) relation obtained for one of the KCl crystal samples. It is clear that this curve is accurately approximated by the expression $q \sim I^2$. Analogous relations were also obtained for NaCl and KBr crystal samples.

An additional test experiment was also performed in which a mirror to reflect emission back to the crystal was used as the sample. In this case the signal dropped significantly, in spite of the increase in the overall light intensity in the test sample. Moreover at $\lambda = 0.27$ μm the photoconductivity of the same samples were measured in a static electric field using the technique described in Chapter 2 of this study. In this case the photoconductivity increases linearally in the intensity range $\sim 10^7 - 10^8$

W/cm^2 and was equal to 10^{-5} - 10^{-4} $Ohms^{-1} \cdot cm^{-1}$, respectively (see Chapter 3).

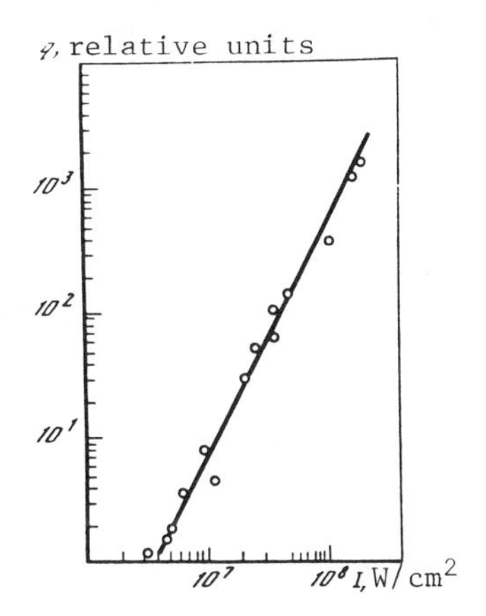

Figure 20. The charge (q) arising across the electrode due to the drag effect on the carriers plotted as a function of laser emission intensity in KCl; $\lambda = 0.27$ μm.

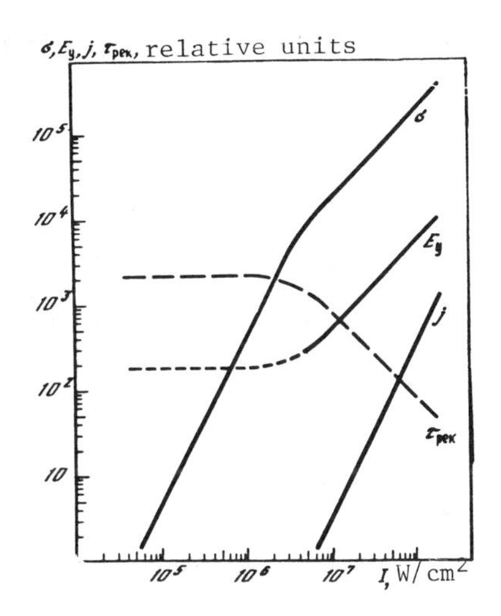

Figure 21. The current and field strength for the drag effect, the photoconductivity and the recombination time plotted against the emission intensity at $\lambda = 0.27$ μm (qualitatively).

Solid line curves correspond to observed relations.

Since the Maxwellian relaxation time (τ_M) exceeds the pulse duration (even with maximum values of the photoconductivity achieved in our experiments, $\sigma \sim 10^{-4}$ $Ohms^{-1} \cdot cm^{-1}$, so that $\tau_M \sim 10^{-8}$ sec $\geqslant \tau_И$), which is a necessary condition to investigate the drag effect and the photoconductivity in dielectrics with capacitive coupling between the sample and the electrodes, the relation plotted in Figure 20 characterizes the drag current density $j(I)$, since $q \sim j$ with a fixed laser beam area. In the given experimental geometry, however, it is not possible to measure the current density with acceptable accuracy. Hence we proposed [86] and implemented a method of direct measurement of the electric field strength of the drag effect (E_y). A pulsed amplitude-controlled voltage was injected to one of the electrodes to compensate the drag effect field. The signal was taken from the second electrode whose potential remained fixed accurate to 10^{-2} V. With equality between the field strengths of the external field (E_{BH}) and the drag effect field, the signal was equal to zero, and when $E_{BH} > E_y$ the signal changed polarity.

It was then possible to measure the voltage of the drag effect. At a laser emission intensity of $\sim 10^8$ W/cm^2 its value was 3 ± 1 V/cm in KCl.

Such a technique also made it possible to measure the $E_y(I)$ relation which was linear for all test crystals.

The drag current density in the short circuit mode is determined [89] by the expression

$$j = K æ e \tau W \hbar q_{ph}/m^*, \tag{9}$$

where τ is the momentum relaxation time of an electron in the dielectrics; W is the photon flux density; $\hbar q_{ph}$ is the photon momentum; K is the dimensionless coefficient determined by the processes occurring in light absorption. Thus, for example, for the drag effect associated with the photoionization of impurity centers [89],

$$K = -(8/5)(\hbar\omega - E_i)/\hbar\omega.$$

Here E_i is the ionization energy of the i-type impurity centers; $\hbar\omega$ as before is the quantum energy at the frequency of the irradiating light.

It is clear that when $K \sim -1$ if we do not select the quantum energy to be close to the ionization energy of the impurity. The "minus" sign infers that the electron flow is in the same direction as the photon flow. For the other mechanisms of the drag effect the coefficient K is different (and may even change sign), although we may assume that $|K| \sim 1$ for rough estimates.

It is not convenient to use formula (9) in an experimental investigation of the drag effect in dielectrics, since it is impossible to provide zero electric field in practice. Hence it is more convenient to use the formula for the open circuit state which is easily achieved in experiment, since the input impedances of the charge-sensitive amplifiers are rather high ($\sim$ 100 kΩ).

In conditions providing a constant drag current and conductivity lengthwise along the dielectric sample, the field strength of the drag effect is $E_y = j/\sigma$, or, accounting for the fact that $\sigma = en\mu$,

$$E_y = jm_T/(e^2 n \tau_T). \tag{10}$$

Here τ_T is the momentum relaxation time of the thermalized carriers (in our experiments $\tau_T \sim 10^{-13} - 10^{-14}$ sec [90]), m_T is the mass of the thermalized carriers.

We should note that we are fully justified in neglecting the field "heating" of the carriers since when $I \leqslant 10^8$ W/cm^2 the additional energy acquired by an electron in the conduction band due to heating ($E_{доп} \sim I$) does not exceed 0.02 eV. This estimate is obtained accounting for the fact that at fields corresponding to intensities where bulk damage to alkali-halide crystals will occur at $\lambda = 0.27$ μm ($\sim 5 \cdot 10^{10}$ W/cm^2 [20]), $E_{доп} \leqslant$ $\leqslant 1$ eV, even if the breakdown mechanism is avalanche ionization [1, 3].

Therefore we take τ_T to be independent of W. Finally for the drag effect field we have

$$E_y = K æ \tau W m_T \hbar q_{ph}/(e\tau_T n m^*). \tag{11}$$

In principle in addition to the drag effect voltage may also arise from other physical processes such as the formation of thermo-EMF due to the nonuniform absorption of light (the Dember effect) or EMF arising from a gradient in the carrier concentration which may occur as light is absorbed by the deep impurity centers. Estimates reveal, however, that the EMF values in these cases are quite low, of the order 10^{-2} - 10^{-3} V, which is significantly lower than the EMF values we obtained in experiment. The formulation and geometry of the experiments described at the beginning of this section also eliminated nonstationary photo-EMF on the crystal surfaces [91], since the laser beam did not make contact with the lateral facets of the samples.

Table 1

Results From an Analysis of the Various Possibilities For Implementation of the Drag Effect For a Number of Physical Processes

Process	$j \sim I^a$	$E_y \sim I^b$	$\sigma \sim I^c$
	a	b	c
Intraband absorption	p + 1	1	p
Transitions to the conduction band for $\tau_{рек}$ = const	p	0 p –	p p –
Transitions to the conduction band for $\tau_{рек} \sim n^{-1}$	p	2	2

Moreover it is possible to propose several processes occurring in the bulk of the test crystals and producing EMF in an open electric circuit. If, for example, the drag effect is attributed to intraband optical transitions, then $æ = n\sigma_0$, where σ_0 is the absorption cross-section of light by an electron in the conduction band. As in [92] assuming $\sigma_0 = 4\pi e\mu/cn'$, where $\mu = 10$ $cm^2 \cdot V^{-1} \cdot sec^{-1}$ [69], n' is the refractive index, we find that $\sigma_0 \sim 10^{-18}$ cm^2. In this case

$$E_y = K\sigma_0 \tau I m_T/(e\tau_T c m^*), \tag{12}$$

where c is the speed of light in the vacuum. Estimates by formula (12) carried out $m^* = m_T$ and $\tau = \tau_T$ show that even at maximum laser emission intensities achieved in experiment, the field strength of the drag effect with intraband transitions is $\sim 10^{-2}$ V/cm, which lies below the detection threshold in our experimental conditions.

If the drag effect is attributed to the transition of electrons to the conduction band, then with a fixed carrier recombination time and remembering that $n \sim I^p$ and $æ \sim I^{p-1}$, we obtain from expression (11) for one- and multi-photon (in the general case) processes

$$E_y = Kp\tau\hbar q_{ph} m_T/(\tau_T e \tau_{рек} m^*), \tag{13}$$

where p is the number of photons involved in the formation of a single electron-hole pair in the ionization process. It is significant that in this case the field of the drag effect with K accuracy is independent of the type of transition during the carrier creation process: Either band-to-band or photoionization of impurities. Estimates based on formula (13) have yielded a value of $E_y \sim 10$ V/cm for $\tau_{рек} = 3 \cdot 10^{-11}$ sec (such a value of the recombination time was obtained for KCl with similar excitation conditions in [37, 38]) and for $m^* = m_T$, $\tau = \tau_T$.

For more accurate calculations we must know the dependence of τ and m^* on the energy of the electrons in the conduction band. We note that expression (13), since it is independent of many of the characteristics of the test substance, represents one of the simplest methods of estimating the carrier recombination time, which is a very complex problem for most dielectrics.

The problem of the value of the recombination time in alkali-halide crystals and particularly the issue of its dependence on the nonequilibrium carrier concentration were until recently disputed and unresolved in solid state physics. Study [25], for example, based on an investigation of photoconductivity in KBr proposed that with σ corresponding to values of $n \geqslant 10^{11}$ cm^{-3} the recombination time becomes independent of n and drops with growth of n as $\tau_{рек} \sim n^{-1}$ which, in the opinion of the authors of [25] reveals square-law carrier recombination at such concentrations. Experiments devoted to the laser photoconductivity of alkali-halide crystals in our study [41] were interpreted analogously.

Moreover the weakening of the $\sigma(I)$ relation from a square-law to a linear law observed in studies [25, 41] (at $I \sim 10^7 - 10^8$ W/cm^2) cannot yet be considered sufficient proof of the fact that this change is attributable specifically to the quadratic recombination of carriers. Thus, for example, we may also allow another interpretation of the derived results that was also considered in [41] where the nature of the change in photoconductivity observed in the high intensity range may be entirely related to the saturation of the impurity energy levels from cascade ionization processes. As will be demonstrated below investigating the characteristics of the drag effect will make it possible to solve the problem of quadratic carrier recombination in alkali-halide crystals.

Results from an analysis of the various possibilities for the implementation of the drag effect are given in Table 1. The table is designed to correlate (in one column) for many physical processes the drag current, the field strength of the drag effect and the photoconductivity as power functions of the laser emission intensity. The power for I is either a constant value (0 or 1) or represents the number of photons involved in a single event of energy and momentum transfer to an electron in the conduction band. Specifically it is clear from Table 1 that if we base our analysis solely on the nature of the relations $j(I)$, $E_y(I)$, $\sigma(I)$, two processes correspond to the experimentally observed curves presented in a single diagram for convenience (Figure 21): Intraband energy absorption for $p = 1$ and the transition of carriers to the conduction band for $\tau_{рек} \sim \sim n^{-1}$ and for $p = 2$. However in the first case as discussed above $E_y \sim \sim 10^{-2}$ V/cm, which is two to three orders of magnitude smaller than the value of E_y we observed experimentally.

These data lead to the conclusion that the drag effect observed in our experiments may be attributed to the transfer of photon momentum to the free carriers in two-photon ionization processes. Bearing in mind that $E_y \sim 2$-5 V/cm (at $I \sim 10^8$ W/cm^2) for the test crystals, we may state that quadratic (square-law) carrier recombination occurs at these excitation levels.

12. Nonlinear Absorption of Laser Radiation at 0.35 and 0.27 μm in Alkali-Halide Crystals

Recently there has been extensive interest in investigating optical emission absorption in ultrapure materials of various types, particularly wideband dielectrics. This interest is related to the need to establish the absorption mechanisms over a very broad frequency range (from the far IR to the UV) in connection with the use of wideband dielectrics as components in high-power laser systems. The investigation of intrinsic absorption mechanisms caused by the excitation of nonequilibrium carriers due to impact or multiphoton ionization plays a special role [8-11].

Moreover the investigation of nonequilibrium carrier generation and recombination mechanisms in wideband dielectrics has been the subject of increasing interest. On the one hand this is related to the special role played by photoelectrons in the radiation defect creation process [37, 38, 47] and, on the other, to the solution of the problems of the critical laser radiation strength of dielectrics [1-7, 12-15, 20, 55]. A number of studies from this brief list are devoted to alkali-halide crystals.

It was demonstrated in the preceding section based on experimental investigations of the drag effect that quadratic recombination processes ensue in alkali-halide crystals when $n \geqslant 10^{12}$ cm^{-3}. Moreover the observed linear character of the dependence of the photoconductivity of alkali-halide crystals on the laser excitation emission intensity in the high intensity range (see Chapter 3 of this study) could be attributed to the appearance of cascade ionization processes through the impurity levels in the bandgap from the saturation of one of the transitions. Investigating light absorption makes it possible to differentiate the quadratic carrier recombination process from two-photon creation (in this case the absorbed energy is $E_{погл} \sim I^2$, while $\sigma \sim I$) from the saturation of one of the transitions in cascade ionization, for which $E_{погл} \sim I$ and $\sigma \sim I$.

With such emission intensities where the quadratic recombination mechanism becomes significant, radiation defects are effectively created towards the end of the laser pulse in alkali-halide crystals [37, 38]. The deep trap centers (recombination centers) that arise from UV irradiation cause a weakening of the $\sigma(I)$ relation due to the retardation in the growth of carrier concentrations in the conduction band. Ionization of the resulting, easily-ionized centers (together with trap centers at the same concentration [38]) may in certain conditions become more probable than electron-hole pair creation as a result of two-photon ionization [93]. The primary contribution to absorption in this case will come from the ionization of recreated centers, since the absorbed energy may be a stronger function of the intensity than in the two-photon process. Thus, information on the change in the effective absorption coefficient is important information in establishing the problem of photoionization mechanisms in wideband dielectrics.

A number of studies have performed extensive investigations of laser emission energy absorption processes in solids by means of optoacoustic techniques. For example [94] proposed a photoacoustic method of detecting absolute values of the thermal characteristics of solids and powders; earlier [18] observed shock waves formed from the laser breakdown of ruby with a temporal resolution of tens of nanoseconds; study [95] recorded acoustic waves arising from the mechanical breakdown of solids, while [39] at λ = 1.06 μm using an analogous technique observed compression waves near the breakdown threshold of KDP and DKDP crystals. The optoacoustic method of measuring optical absorption is discussed in more comprehensive and consistent detail in [96] where periodic sequences of laser pulses are used to irradiate the test crystals.

A detailed theoretical analysis of the photoacoustic effect in solids is given in [97] which, specifically, predicts a dependence of the acoustic signal level on the absorption coefficient. Experimental evidence in favor of the hypothesis that in a solid the observed acoustic signal is proportional to the absorbed energy may be found in, for example, study [98].

In [99, 100] we attempted to investigate laser emission absorption in alkali-halide crystals in the UV (at 0.27 and 0.35 μm, respectively), using the acoustooptic method of recording acoustic oscillations. This study also investigated crystals whose photoconductivity we had previously examined at the same wavelengths [41]. This allowed us to properly correlate the research results and to reliably interpret the nonequilibrium carrier photoexcitation processes.

The laser assembly described in Chapter 2 of this study was used in experiments to investigate absorption. The test samples were undoped NaCl, KCl and KBr monocrystals in the form of parallelepiped 20x20x30 mm^3. The samples were freshly cleaved along the cleave planes and were pretested on a spectrophotometer in order to determine the linear absorption coefficient at the test wavelengths.

The emission was focused into the bulk of the samples by means of a long-focus silica lens with a focal distance $f_o \sim 500$ mm. The laser beam diameter was measured prior to the samples by the method proposed in [87, 88] and was 1 mm at $1/e$ of the maximum axial intensity. A diaphragm 1 mm in diameter was placed directly in front of the entrance surface of the samples to avoid parasitic illumination of the crystals. The emission intensities were less than 10^8 W/cm^2, i.e., less than the breakdown threshold of the crystal surfaces [53]. After each laser pulse the condition of the entrance surface of the samples was checked. The absence of a plasma on the entrance surface at high emission intensities was also verified. The lower limit on emission intensity was limited by the sensitivity of the recording equipment to the acoustic signal and was $\sim 10^5$ W/cm^2. The configuration of the assembly used to investigate the absorption of UV laser emission by the optoacoustic technique is shown in Figure 22.

A sensor fabricated from brand TsTS-19 piezoceramics which has one of the highest piezoelectric effect values among all ceramic materials was used as the acoustic sensor. This sensor was fabricated as a cylinder with a diameter of 2 mm and a height of 2 mm and was attached to the surface of the samples by means of a sound-absorbing material for best possible acoustic matching. The sensor was mounted on the lateral surface of the crystals approximately 10 mm from their entrance surface. The distance from the laser beam axis to the piezosensor was 10 mm. With this configuration the signal resulting from the absorption of emission by the entrance surface (due to the unavoidable surface defect layer, impurites, etc. [53]) will be fixed later than the signal attributed to the energy absorbed as a result of irradiation of the crystal bulk. Measures were taken to significantly reduce the influence of parasitic effects due to recording of acoustic oscillations from scattered emission as well as the intrinsic pyroactivity of the sensor. For this purpose aluminum foil 0.05 mm in thickness which is a good reflector of UV was placed between the piezoceramic and the sample. A thin glycerin layer applied to the lateral surface of this sample at the point where it makes contact with the sensor provided a good acoustic match. The scheme explaining the experimental geometry for optoacoustic absorption measurements is shown in Figure 23.

The electric signal from the piezosensor was amplified by means of an electronic amplifier based on the 574UD1-A IC (whose sensitivity is better than 10^{-4} V) and was injected to one of the channels of a S8-2 dual-trace storage oscilloscope. The second channel was designed for energy measurements.

As the laser emission energy is absorbed by the crystal bulk, the illuminated region rapidly heats up. The resulting thermoelastic stresses become the source of compressional waves with a period $d/2v_{3B}$, where v_{3B} is the velocity of sound propagation in the crystals; d is the laser beam diameter. Since the laser beam is axially symmetric within the test samples and due to the relatively low linear absorption coefficient in the alkali-halide crystals at the wavelength used ($\leqslant 10^{-1}$ cm^{-1} [57, 79]), the resulting elastic vibrational wave has cylindrical symmetry.

We may assume in our conditions for integral measurements of the acoustic signal amplitude (with a characteristic relaxation time of the piezosensor of $\sim 10^{-6}$ sec) that the pressure perceived by the sensor is proportional to the acoustic wave pressure on the surface of the sample. This in turn is proportional to the mean pressure (P) within the illuminated region, since in a cylindrical acoustic wave pressure diminishes with distance (r) as $Pr^{-1/2}$.

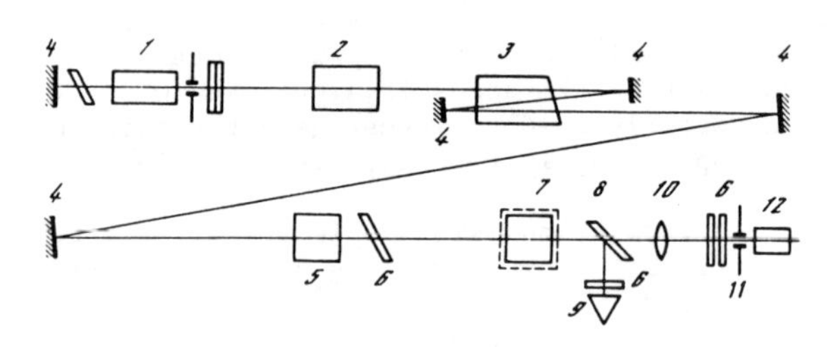

Figure 22. Optical configuration for investigating laser emission absorption by the optoacoustic technique.

1 - YAG:Nd^{3+} laser; 2 - YAG:Nd^{3+} amplifier; 3 - three-pass neodymium glass amplifier; 4 - mirrors; 5 - KDP crystal frequency doubler; 6 - light filters; 7 - DKDP crystal doubler or KDP crystal for obtaining the third harmonic; 8 - beamsplitting silica plate; 9 - pyrodetector; 10 - silica lens, $f_0 \sim 500$ mm; 11 - diaphragm, $d \sim 1$ mm; 12 - sample.

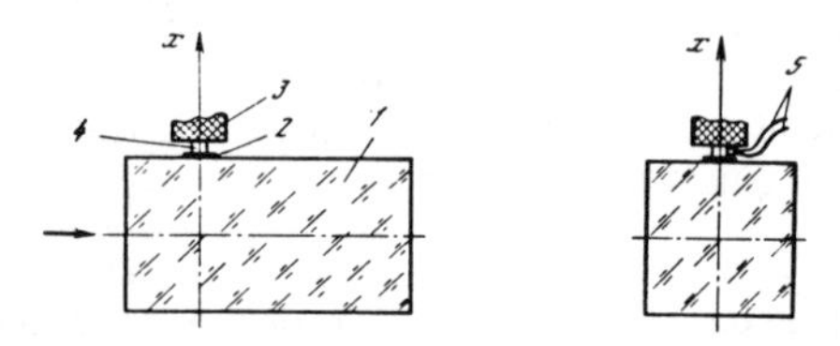

Figure 23. Relative configuration of piezosensor and sample.

1 - Test dielectric crystal sample; 2 - thin aluminum foil; 3 - sound-absorbing material; 4 - piezoceramic sensor; 5 - electric signal amplifier pins; x - axis of symmetry of the sensor. The arrow indicates the direction of laser emission propagation.

Since over the laser pulse duration (remembering that $\tau_{И} \sim 8 \cdot 10^{-9}$ sec) the heat from the irradiated region may spread due to thermal conduction of the material over a characteristic distance $l_T = (\tau_{И}\chi)^{1/2} \sim 10^{-5}$ cm (χ is the thermal diffusivity of the material; for alkali-halide crystals $\chi \sim 4 \cdot 10^{-2}$ cm^2/sec [58]) and since $l_T \ll d$ (we assume for the estimate that the laser beam diameter is constant at 0.1 cm along the sample length), the absolute pressure created by the acoustic wave in the sample bulk may be estimated from the expression $P = \Gamma E_{ПОГЛ}$, where $\Gamma = \gamma v_{ЗВ}^2/C_v$ is Grüneisen's coefficient, γ is the temperature coefficient of volumetric expansion of the material, C_v is the thermal capacity of the test material at constant volume.

Since even at maximum emission intensities used in our case of $E_{ПОГЛ} \sim$ $\sim$ 0.1 J/cm^3 over the pulse time the estimate will produce a value of $P \sim$ $\sim$ 3 bar. At acoustic wave intensities corresponding to such an internal pressure level the natural absorption of sound is significantly linear. Changes in the density of the medium and the propagation velocities of the acoustic waves with these values of P are quite small and may be neglected [101]. The thermal capacity of alkali-halide crystals at such a pressure also varies insignificantly [102].

Thus, in our experimental conditions we need not anticipate nonlinear distortions caused by the transmission of acoustic wave energy from the crystal bulk to the piezosensor whose sensitivity in this case is independent of the value of P across the entire range of laser emission intensities.

In principle the acoustic pressure P may include the acoustic pressure that arises within the crystal due to electrostriction:

$$P_{стр} = \rho(d\varepsilon/d\rho)E_{эфф}^2/8\pi, \qquad (14)$$

where ρ is the density of the medium, ε is the dielectric constant of the medium, $E_{эфф}$ is the effective electric field strength in the laser beam. An estimate of the influence of electrostriction on P obtained from expression (14) for $I \sim 10^8$ W/cm^2 which corresponds to an effective electric field strength of the light wave of 10^4 V/cm shows that $P_{стр} \sim 5 \cdot 10^{-5}$ bar, i.e., it is small and may be ignored.

The acoustic signal caused by energy absorption in the bulk of the crystals appeared as damped complex oscillations of opposite polarity whose generation was delayed with respect to the initiation of the laser pulse by a time corresponding to the propagation time of the acoustic wave from the laser beam axis to the piezosensor. A typical oscillogram of the acoustic signal is shown in Figure 24 for one of the NaCl crystal samples. The amplitude of the first positive half-cycle was used as the test acoustic signal.

The observed change in the time gap from the generation of the laser pulse to the excitation of the acoustic oscillations with a change in the beam position with respect to the samples corresponded to the beam propagation time from the illuminated bulk to the piezosensor. For a fixed laser beam position with respect to the samples, the delay of the onset of acoustic oscillations varied for different alkali-halide crystals, corresponding to the acoustic velocities in them (for example, the ratio of acoustic velocities is ~ 1.3-1.4 for NaCl and KBr [81]).

Irradiating the samples by laser emission at λ = 1.06 μm of the same intensity as in the UV did not produce an acoustic signal of significant amplitude, since the test alkali-halide crystals at λ = 1.06 μm have very low linear absorption coefficients: $\varkappa_1 \sim 10^{-4}$ cm^{-1} [57]. In order to investigate the nature of the dependence of absorbed energy on laser emission intensity it was not necessary to know the absolute values of the absorbed energy, since the piezosensor was not calibrated. The accuracy of relative amplitude measurements of the acoustic signals was 10%.

Figure 25, a, shows the dependence of acoustic signal amplitude (A) on laser emission intensity at λ = 0.35 μm obtained for three KBr crystals having different linear absorption coefficients at this wavelength. The quantities $\varkappa_1$ for these KBr samples obtained by measuring absorption on a spectrophotometer are given in Table 2. The relation (Figure 25, a) shows two characteristic sections: A linear section in the relatively low emission intensity range ($3 \cdot 10^5 - 8 \cdot 10^6$ W/cm^2) and a quadratic (square-law) section at I ~ $10^7 - 10^8$ W/cm. It is clear that in spite of the significant difference in the values of $\varkappa_1$ (greater than a factor of three) absorption becomes virtually identical in the high intensity region for these samples. Analogous A(I) relations were also observed for other KBr samples.

The A(I) relation at λ = 0.27 μm obtained for two KCl crystal samples having different values of $\varkappa_1$ (see Table 2) is shown in Figure 25, b. Here we also observe two characteristic sections: A linear section for I ~ $10^5 - 10^7$ W/cm^2 and a quadratic section for I $\geqslant 10^7$ W/cm^2. Variable absorption by the samples in the first section (varying by a factor of 2-2.5 if we base the estimate on the acoustic signal) is replaced by virtually identical absorption in the high intensity range. An analogous behavior was observed at λ = 0.27 μm for NaCl samples with different $\varkappa_1$.

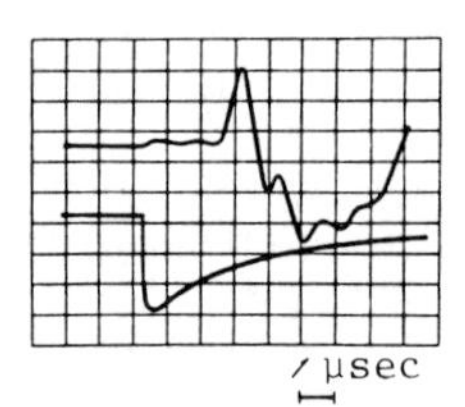

Figure 24. Oscillogram of an acoustic signal in NaCl (upper waveform); lower waveform: Integrated laser pulse; λ = 0.27 μm.

Figure 26 gives the A(I) relations obtained at λ = 0.27 μm for samples of different alkali-halide crystals having similar linear absorption coefficients at this wavelength. With relatively low emission intensities the curves virtually merge in the linear section of the relations (for NaCl and KCl crystal samples in Figure 26, a, and KBr and KCl crystal samples in Figure 26, b), while in the high intensity region they noticeably diverge with an identical angle of inclination corresponding to a square-law A(I) relation. This indicates a significantly different observed magnitude of nonlinear absorption in the various alkali-halide crystals.

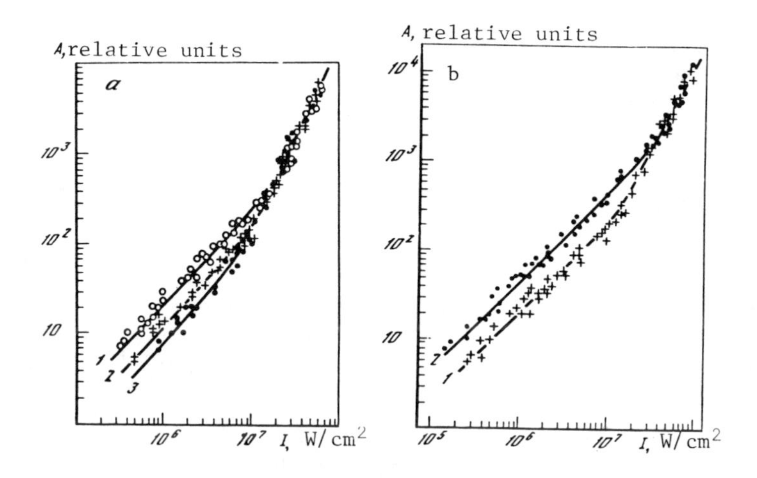

Figure 25. Acoustic signal amplitude plotted as a function of laser emission intensity for three KBr samples (a) and two KCl samples (b) having different linear absorption coefficients.

a: $\varkappa_1$ = 0.12 (1); 0.07 (2); 0.04 cm^{-1} (3); b: $\varkappa_1$ = = 0.08 (1); 0.24 cm^{-1} (2); λ = 0.27 μm.

Since for all the test crystals the A(I) relations have two sections (a linear section and a quadratic section), these sections naturally are related to the different physical mechanisms according to which energy is absorbed. The existence of a linear section indicates that in the relatively low intensity region absorption clearly related to ionization of the impurity states predominates. The appearance of a quadratic section in the high intensity range may only be linked to the fact that nonlinear absorption exceeds linear absorption, so that the absorption coefficient in this case becomes proportional to I. This case will in fact exist in two-photon (band-to-band) energy absorption.

We should emphasize that we investigated media with similar acoustic parameters to those of alkali-halide crystals [58, 81] although these samples in a number of cases had significantly different linear absorption coefficients. In this connection the fact that in the linear section of A(I) the strongest acoustic signal is always observed in samples having the highest value of $\varkappa_1$ which reveals an impurity source of absorption.

Table 2

Linear Absorption Coefficients (in cm^{-1}) For a Number of Alkali-Halide Crystal Samples Obtained by Spectrophotometer Absorption Measurements

Sample	λ = 0.35 μm	λ = 0.27 μm	Sample	λ = 0.35 μm	λ = 0.27 μm
NaCl		0.23			
KBr - 1	0.12	-	KBr - 4	-	0.07
KBr - 2	0.07	-	KCl - 1	-	0.08
KBr - 3	0.04	-	KCl - 2	-	0.24

Moreover as we see from Figure 25 in the high intensity range all curves for samples of the same alkali-halide crystal type yet with different $æ_l$ values virtually coincide, which indicates only one possible explanation: The impurity nature of absorption ceases to dominate. The transition from the linear to the square-law dependence of the acoustic signal amplitude on laser emission intensity also reveals a change in the energy absorption mechanism. The fact that with a common quadratic A(I) relation in different alkali-halide crystals (see Figure 26) a different nonlinear absorption value is observed characteristic of each of the crystals indicates an intrinsic absorption mechanism.

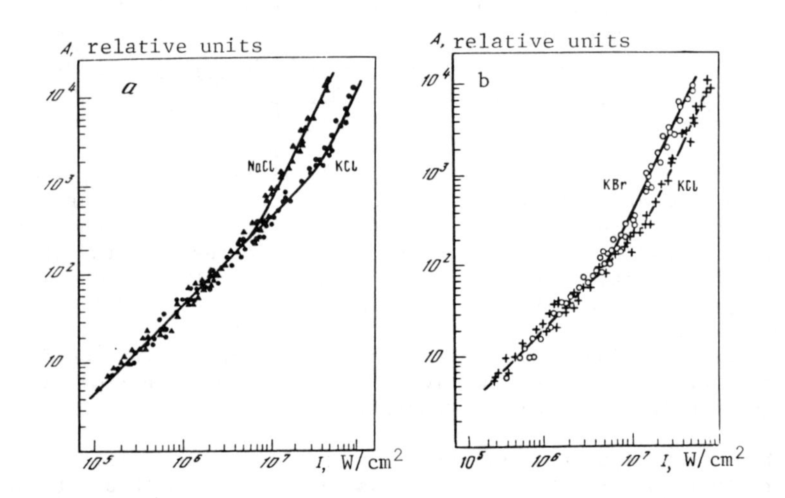

Figure 26. The A(I) relation for different crystal samples.
a: NaCl, $æ_l$ = 0.23 cm^{-1}, KCl, $æ_l$ = 0.24 cm^{-1}; b: KBr, $æ_l$ = = 0.07 cm^{-1}, KCl, $æ_l$ = 0.08 cm^{-1}; λ = 0.27 μm.

On the other hand we know that with two-photon carrier generation a significant portion of the absorbed energy enters the lattice (here $E_{\text{погл}} \sim \sim I^2$), so that at high excitation levels this process manifests an intrinsic nature and is independent of the individual properties of the samples associated with the existence of various types of impurities in them. These data from acoustooptic measurements therefore reveal that in our experimental conditions two-photon (band-to-band) energy absorption is realized.

Table 3

Two-Photon Absorption Coefficients (in 10^{-3} cm/MW) For a Number of Alkali-Halide Crystal Samples Measured by the Optoacoustic Technique

Sample	λ = 0.35 μm	λ = 0.27 μm	Sample	λ = 0.35 μm	λ = 0.27 μm
NaCl		9.2			
KBr - 1	3.1	-	KBr - 4	-	3.5
KBr - 2	2.4	-	KCl - 1	-	2.1
KBr - 3	1.7	-	KCl - 2	-	4.0

Since the square-law sections of the A(I) relations do not coincide for different alkali-halide crystal samples even with approximately equal initial absorption, we may estimate the two-photon absorption coefficient. We carried out the estimate by fitting the A(I) relation observed in experiment to a separately measured linear absorption coefficient. In this case it was assumed that the value of $\varkappa_1$ for each of the samples was constant over the entire laser emission intensity range. Moreover, we assume that the majority of absorbed energy is transferred to the lattice in a time period less than the response time of the piezosensor (~ 1 μsec) in both one-photon and two-photon processes.

Table 3 presents the values of the two-photon absorption coefficients estimated with these assumptions ($\varkappa_2$) for a number of alkali-halide crystal samples. We note that these quantities are similar to the values measured in [49] by light transmission at these same wavelengths.

The values of $\varkappa_2$ for the different samples of the same alkali-halide crystal were somewhat different evidently due to the fact that not all of the absorbed laser emission energy was liberated as heat over the short time period. Some of the absorbed energy is expended in exciting luminescence centers and "long-lived" centers and in conserving luminescence for a period exceeding the period of the acoustic oscillations. It is clear that the portion of energy absorbed as heat which is also attributable to the acoustic signal may vary from sample to sample depending on the concentration of the various impurities and defects.

The data given in Chapter 3 of this study on the photoconductivity of alkali-halide crystals with a transition from a square-law $\sigma(I)$ relation to a linear relation in the high emission intensity range of $I \geqslant 5 \cdot 10^6$ W/cm^2 make it possible to conclude from a comparison to results from opto-acoustic investigations of the nonlinear absorption of the same crystals at the same wavelengths and to results from an investigation of the photon drag effect on carriers in the UV that photoconductivity in this range of intensities is attributable to two-photon band-to-band ionization. The transition of the (I) relation from a square-law to a linear law in this case may be explained only by the dominant influence of quadratic recombination.

CHAPTER 5

KINETIC MODEL OF THE PHOTOCONDUCTIVITY OBSERVED IN ALKALI-HALIDE CRYSTALS

This chapter presents a brief theoretical analysis of the observed regularities of photoconductivity excited in alkali-halide crystals under UV laser irradiation (at 0.27 and 0.35 μm). This analysis was carried out using a kinetic model that we believe was the most advantageous model that includes single-photon ionization processes of the impurities as well as two-photon ionization of the crystalline lattice atoms, linear and quadratic carrier recombination and radiation defect creation.

13. The Role of Defects in Observed Laser Photoconductivity Phenomena and Analysis of Kinetic Equations

Recently the formation of radiation defects in alkali-halide crystals under UV laser irradiation has been the subject of intense research (both experimental and theoretical) [37, 38, 47]. Their formation mechanism is briefly as follows. The holes created from two-photon ionization over times $\sim 10^{-13}$ sec are self-localized, forming Cl_2^- configurations with neighboring atoms (halogens) and then trapping the free electrons, forming self-trapped Cl_2^{2-} excitons in one of the excited states. This opens a quadratic carrier recombination channel. The lifetime of electrons in the conduction band in this case is determined by $(\alpha n_h)^{-1}$, where n_h is the concentration of self-localized holes, α is the coefficient determining the rate of the quadratic carrier recombination process. The self-trapped exciton then either relaxes to the lower energy states from radiative recombination or decays from the excited state into a pair of radiation defects: F- and H-centers. This decay occurs over times $\sim 10^{-11}$ sec.

We note that in the previously cited studies [37, 38] which employed piezosecond pulses the peak in concentration of radiation defects were observed after the end of the laser pulse. Defect pairs with a minimum distance between them recombine over times $\sim 10^{-12}$ sec; more remote defect pairs may be conserved for a long time period, since their recombination involves their thermal motion. Therefore in [38] the efficiency of defect creation over 10^{-11} sec was $\eta \sim 0.3$ defect pairs/electron hole pairs, and was ~ 0.1 of the initial defect quantity 10 sec after the end of the laser pulse. We may conclude on this basis that the recombination processes develop primarily on a nanosecond time scale.

As indicated by our experiments the direct ionization of F-centers arising during the UV irradiation of the crystals is significant for the photoconductivity signal. Moreover, ionization of the F-centers opens the familiar stable center formation channels (such as the F_A-centers) at the same time hindering the recombination of F-H pairs. Therefore the yield of F-centers when $\tau_И \sim 10^{-9}$ sec may be significantly greater than when $\tau_И \sim 10^{-11}$ sec. This is confirmed by the fact that even at $I \sim 10^7$ W/cm^2 the concentration of F-centers is 10^{16} cm^{-3}. These centers are stable in darkness, i.e., they do not recombine with H-centers as a result of thermal motion.

By illuminating the samples with a powerful halogen lamp in the absorption band of the F-centers the initial color centers are destroyed, although a process such as the photoexcitation of F-centers and thermal ionization of electrons with electron trapping by the H-centers will not result in defect recombination but rather will result in the formation of F^+ (vacancy) and H^--center pairs. Of course we cannot, generally speaking, eliminate the photostimulation of radiation defect recombination as a result of, for example, an increase in their mobility (their breakaway from the impurities, dislocations, etc.), although clearly these processes play a secondary role.

Our primary assumption is that regardless of whether or not the experiment is conducted with additional illumination to eliminate the initial F-centers the irreversible increase in photoconductivity is provided by electrons belonging to defects that are reactivated under UV irradiation.

With this assumption the rigid correlation between the form of the $\sigma(I)$ relation is the second linear section and the irreversible increase in photoconductivity is a natural process that is, moreover, a necessary process.

At the same time the portion of photoelectrons involved in defect formation is significant, when the quadratic (related to the formation of exciton states) recombination significantly exceeds the linear recombination time. Moreover the rate of growth in the number of defects is porportional to n^2, i.e., I^4 on the quadratic section of $\sigma(I)$. Thus we may understand the virtual impossibility of observing accumulation effects and the onset of an irreversible growth in photoconductivity in the low intensity range.

We will consider the possible role of certain simple and important defects in the formation of the $\sigma(I)$ relation. The vacancies at room temperature play the role of "shallow" traps (with a depth of the order of kT). Indeed electron trapping by the vacancies produces F*: Excited states of the F-centers. The relaxation time of the F*-centers to the ground state is $\sim 10^{-6} - 10^{-7}$ sec $\gg \tau_И$ at the same time that the thermal ionization time of the electrons at room temperatures $\geqslant 10^{-10}$ sec.

The "positively charged" H-centers may function as effective deep trap centers by transforming into H^{-}-centers. Unfortunately at present there are no reliable data indicating stability and thermal mobility of such centers. Aggregate defects of the M-, N-center, F_A-center, etc., type arise from the formation of photochemical by-products (these defects are in some sense analogs of the F-centers: They have excited states with an excess electron, etc.

The primary purpose of this analysis, however, is to systematize the model representation to provide a correct $\sigma(I)$ relation. Henceforth we shall therefore assume that in the irradiation of the crystals by UV laser irradiation there generally arise two types of defects: Ionization centers (in the n_1 concentration) and trap centers (in the n_2 concentration).

For simplicity we will neglect the exciton decay time into F-H-pairs. We will write a system of kinetic equations valid for the pulse duration:

$$\begin{aligned} dn/dt &= \delta\xi^2 + \varepsilon_0 n_0 \xi + \varepsilon_1 n_1 \xi - \alpha n n_h - \beta_0 n_{r_0} n - \beta_2 n_2 n, \\ dn_h/dt &= \delta\xi^2 - \alpha n n_h, \\ dn_1/dt &= \eta \alpha n n_h - \varepsilon_1 n_1 \xi, \\ dn_2/dt &= \eta \alpha n n_h - \beta_2 n_2 n. \end{aligned} \qquad (15)$$

The first equation describes the change in free electron concentration with time. Here ξ is the dimensionless emission intensity: $\xi = I/I_o \sim$ $\sim 5 \cdot 10^6$ W/cm^2; $\delta\xi^2$ is the number of electron-hole pairs created per unit of time per unit of crystal volume as a result of two-photon ionization (δ is the kinetic coefficient); $\varepsilon_o n_o \xi$ is the same number of pairs for the single-photon ionization of impurities or defects existing prior to high-power UV irradiation (n_o is their concentration, ε_o is the kinetic coefficient); $\varepsilon_1 n_1 \xi$ is the number of free electrons created by single-photon ionization from the recreated defects (ε_1 is the kinetic coefficient); the term $\alpha n n_h$ accounts for the quadratic recombination of carriers, while the term $\beta_o n_{ro} n$ and $\beta_2 n_2 n$ represent the linear recombination process by centers existing in the crystal prior to high-power UV irradiation and by the recreated centers, respectively (n_{ro} is the concentration of "empty" traps, β_o and β_2 are the kinetic coefficients).

The second equation describes the change in hole concentration from two-photon ionization and quadratic recombination. The third and fourth equations describe the dynamics of processes occurring at the recreated centers (ionization and trapping, respectively) ($\eta < 1$ is the coefficient accounting for the recombination of the centers themselves).

Kinetic equation system (15) has no analytic solution. Here we will limit our examination to a qualitative consideration of the model determined by system (15).

We will briefly describe the physical processes responsible for the nature of the $n(t)$ and $n(\xi)$ relations. When the inequality

$$\sqrt{\delta/\alpha}/(\varepsilon_0 n_0/(\beta_0 n_{r_0}) > 1 \qquad (16)$$

is satisfied in the $\xi < 1$ range the nature of the $n(t)$ function is shown in Figure 27, a. With forward progression of the $n(\xi)$ relation over time $t_1 \sim 1/\beta_o n_{ro}$ the $n(t)$ function reaches its peak: $n_{max} = (\delta\xi^2 + \varepsilon_o n_o \xi)/\beta_o n_{ro}$. It is clear that in this case the dominant role is played by single-photon impurity ionization processes and two-photon carrier creation processes, while the rise time to the peak is determined by the linear recombination process. Then $n(t)$ decays very slowly ($dn\ dt \sim -(10^{-2} - 10^{-5})$) to the steady-state value $n_\infty = \varepsilon_o n_o \xi/\beta_o n_{ro}$ and will come closer to this value the more the pulse duration exceeds the linear recombination time $\tau_И > \tau_{рек} = 1/\beta_o n_{ro}$. This decay is caused by the onset of the quadratic carrier recombination process. In fact $\tau_И$ is located on the quasistationary segment of the curve and in experiment a photoconductivity signal corresponding to n_{max} is measured.

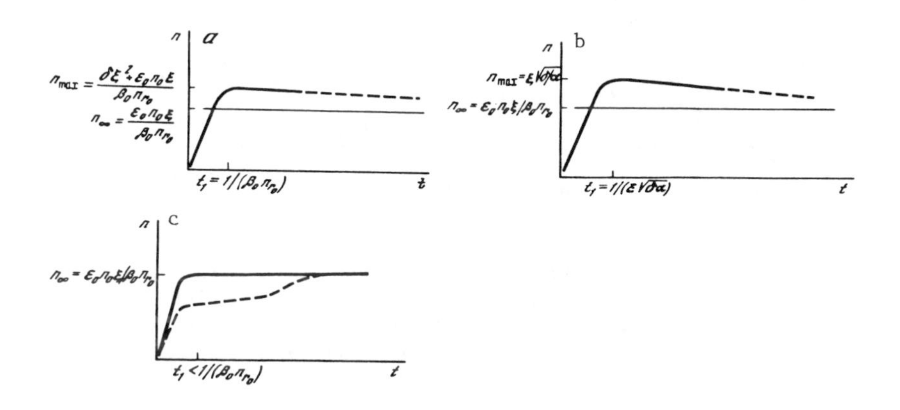

Figure 27. The change in nonequilibrium carrier concentration (n) with time accounting for their linear recombination.

a - Low intensity range; b - high intensity range; c - auxiliary case.

In the $\xi > 1$ range the nature of the $n(t)$ relation is the same; only the value of $n_{max} = \xi\sqrt{\delta/\alpha}$ changes: Over a characteristic time $t_1 = (\xi\sqrt{\delta\alpha})^{-1}$ the $n(t)$ function reaches its peak (Figure 27, b) corresponding to the electron concentration in the second linear section of $n(\xi)$. This rise process is determined by quadratic recombination. Then $n(t)$ approaches the steady-state value n_∞ slowly (due to linear recombination). As before in reality $n \sim n_{max}$ is recorded.

When an inequality opposite that of (16) is satisfied the nature of the $n(t)$ relation is different (Figure 27, c): Over time $t_1 < (\beta_o n_{ro})^{-1}$ the function rapidly approaches its stationary value. In principle in this case the change in the $n(t)$ function will become more complex (represented by the dotted line) due to the realization of the following mechanism: 1) The onset of quasistationarity due to quadratic recombination; 2) a slow growth of n due to single-photon ionization of the impurities, which in turn causes a reduction in the lifetime and concentration of the holes and a further increase in n, etc. The linear recombination represents the "competition" of this process whereby the $n(t)$ function reaches its stationary value of $n_\infty = \xi\varepsilon_0 n_0/\beta_0 n_{ro}$. It is clear that the $n(\xi)$ relation in this case is linear.

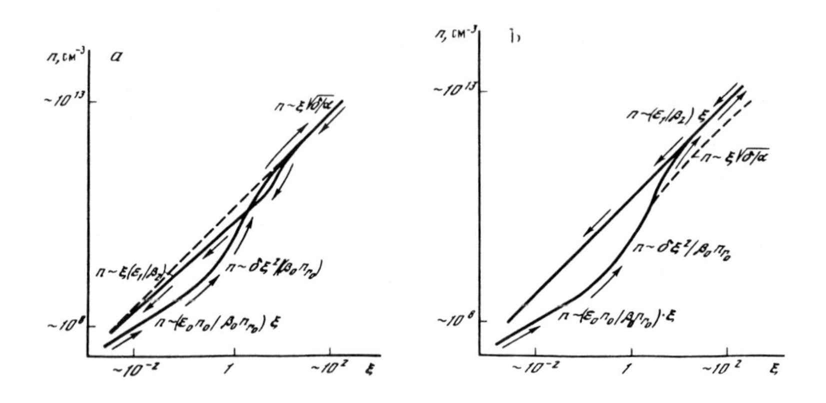

Figure 28. The nonequilibrium carrier concentration (n) plotted as a function of UV laser emission intensity (ξ) accounting for ionization and trap center creation processes (qualitatively) both when inequality (18) is satisfied (a) and not satisfied (b).

The arrows indicate the direction of change in intensity, while the dashed line represents the unrealized position of $n(\xi)$.

According to this model in addition to the described processes two types of defects are also created: Ionization and trap centers and therefore in the second linear section of $n(\xi)$, i.e., with high values of ξ the quantities change: $\varepsilon_0 n_0 - \varepsilon_0 n_0 + \varepsilon_1 n_1$, $\beta_0 n_{ro} - \beta_0 n_{ro} + \beta_2 n_2$. Since the concentrations of the created defects are rather large:

$$n_1 = n_2 \sim \delta\xi^2\tau_H \gtrsim 10^{15}\ \text{cm}^{-3} \quad \text{when} \quad \xi \sim 10, \text{ then} \quad \varepsilon_1 n_1 \gg \varepsilon_0 n_0 \quad \text{and} \quad \beta_2 n_2 \gg \beta_0 n_{r_0}.$$

In this case the ratio of kinetic coefficients determining n from the combined single-photon ionization and linear recombination processes, becomes

$$(\varepsilon_0 n_0 + \varepsilon_1 n_1)/(\beta_0 n_{r_0} + \beta_2 n_2) \approx \varepsilon_1/\beta_2. \tag{17}$$

We note that relation (17) is independent of the concentrations of created defects and determines the slope of the linear $n(\xi)$ relation in its reverse progression. Thus, it is necessary for the inequality $(\varepsilon_1/\beta_2) >$ $> (\varepsilon_o n_o/\beta_o n_{ro})$ to be satisfied in order to create the observed irreversible increase in the photoconductivity of the crystals, where the $\varepsilon_o n_o/\beta_o n_{ro}$ relation determines the slope of the first linear section of the $n(\xi)$ relation during its forward progression. Here two cases are possible.

Assume the following inequality is satisfied after the formation of the new centers (the ionization and trap centers)

$$\sqrt{\delta/\alpha}/(\varepsilon_1/\beta_2) > 1. \qquad (18)$$

Then there will normally be three sections in the forward progression of $n(\xi)$: A linear section, a quadratic section and another linear section, while $n \sim \xi\sqrt{\delta/\alpha}$ in the second linear section (Figure 28, a). In the reverse progression $n \sim \xi\varepsilon_1/\beta_2$, i.e., it is determined by the recreated defects.

When the inequality inverse to (18) is satisfied the nature of the $n(\xi)$ relation in principle is identical for the forward and reverse progressions, although in the forward progression the function $n(\xi)$ does not become $n \sim \xi\sqrt{\delta/\alpha}$ but rather becomes $n \sim \xi\varepsilon_1/\beta_2$ (Figure 28, b). The reverse progression of the $n(\xi)$ relation, as in the preceding case, is entirely determined by the recreated ionization and trap centers.

A further growth in emission intensity, although it results in an increase in the concentration of both types of defects will not change the value of the ε_1/β_2 relation, since the slope of $n(\xi)$ remains unchanged, in agreement with experiment.

It is important to emphasize that when $\varepsilon_1/\beta_2 > \sqrt{\delta/\alpha}$ a situation occurs where at high intensities the photoconductivity signal is determined precisely by the single-photon ionization process, although the number of carriers created by two-photon ionization is significantly greater $(\delta\xi^2 \gg$ $\gg \varepsilon_1 n_1 \xi)$. This fact explains the extreme slope of the linear $n(\xi)$ relation in the $\xi \gg 1$ range and the absence of a sublinear $n(\xi)$ relation in the reverse progression. The absorbed laser emission energy in this case is proportional to ξ^2, which is in agreement with optoacoustic measurements.

BIBLIOGRAPHY

1. Epifanov, A. S. The Development of Avalanche Ionization in Solid Transparent Dielectrics Under High-Power Laser Emission Pulses. ZhETF, 1974, Vol. 67, Issue 5, pp. 1805-1817.

2. Zakharov, S. I. The Avalanche Ionization in Transparent Dielectrics Under Near-Breakdown Optical Irradiation. ZhETF, 1975, Vol. 68, Issue 6, pp. 2167-2176.

3. Epifanov, A. S., Manenkov, A. A., Prokhorov, A. M. The Frequency and Temperature Dependencies of Avalanche Ionization in Solids Under Electromagnetic Field Action. Pis'ma v ZhETF, 1976, Vol. 70, Issue 2, pp. 728-737.

4. Manenkov, A. A. New Results on Avalanche Ionization as a Laser Damage Mechanism in Transparent Solids. Laser Induced Damage in Optical Materials, 1977. Wash.: US Government printing office, 1978, pp. 455-464. (NBS spec. publ. 509).

5. Liu, P., Yen, R., Bloembergen, N. Dielectric Breakdown Threshold, Two-Photon Absorption and Other Optical Damage Mechanisms in Diamond. IEEE J. Quant. Electron, 1978, Vol. QE-14, No. 8, pp. 574-576.

6. Gorshkov, B. G., Danileyko, Yu. K., Epifanov, A. S. et al. The Laser Breakdown Mechanism of Alkali-Halide Crystals: An Investigation of the Temperature Dependence of Breakdown Thresholds. Pis'ma v ZhTF, 1976, Vol. 2, Issue 6, pp. 284-287.

7. Gorshkov, B. G., Danileyko, Yu. K., Epifanov, A. S. et al. Laser-Induced Damage to Alkali-Halide Crystals. ZhETF, 1977, Vol. 72, Issue 3, pp. 1171-1181.

8. Braeunlich, P., Schmid, A., Kelly, P. Contribution of Multiphoton Absorption to Laser Intrinsic Damage in NaCl. Appl. Phys. Lett., 1975, Vol. 26, pp. 150-153.

9. Bloembergen, N. Laser-Induced Electric Breakdown in Solids. IEEE J. Quant. Electron, 1975, Vol. QE-110, No. 3, pp. 375-386.

10. Braeunlich, P., Schmid, A., Kelly, P. Starting Times of Laser-Induced Intrinsic Damage in NaCl. Appl. Phys. Lett., 1975, Vol. 26, No. 5, pp. 223-226.

11. Boling, N. L., Braeunlich, P., Schmid, A., Kelly, P. Statistics in Laser Induced Dielectric Breakdown. Appl. Phys. Lett., 1975, Vol. 27, No. 3, pp. 191-194.

12. Gomelauri, G. V., Epifanov, A. S., Manenkov, A. A., Prokhorov, A. M. The Statistical Features of the Avalanche Ionization of Broadband Dielectrics Induced by Laser Irradiation With a Trap Electron Deficit. ZhETF, 1980, Vol. 79, Issue 6, pp. 2356-2363.

13. Gorshkov, B. G., Danileyko, Yu. K., Manenkov, A. A. et al. The Dimensional Effect and Statistics of Laser-Induced Damage in Alkali-Halide Crystals at 10.6 μm. Kvantovaya elektron, 1981, Vol. 8, Issue 1, pp. 148-153.

14. Gorshkov, B. G., Danileyko, Yu. K., Epifanov, A. S. et al. The Influence of UV Irradiation on the Breakdown of Alkali-Halide Crystals by CO_2 Laser Emission. Kvantovaya elektron, 1981, Vol. 8, Issue 1, pp. 155-156.

15. Gorshkov, B. G., Epifanov, A. S., Manenkov, A. A., Panov, A. A. The Breakdown of Optical Materials in Crossed Laser Beams at Various Frequencies. Izv. AN SSSR. Ser. fiz., 1980, Vol. 44, Issue 10, pp. 2062-2065.

16. Byob, R. Fotoprovodimost' tverdykh tel [The Photoconductivity of Solids]. Moscow: Izd-vo inostr. lit., 1962, 558 pages.

17. Belikova, T. P., Sviridenkov, E. A. The Photoconductivity of Ruby Under High-Power Irradiation by a Ruby Laser. Pis'ma v ZhETF, 1966, Vol. 3, Issue 10, pp. 394-398.

18. Belikova, T. P., Sovtsenko, A. N., Sviridenkov, E. A. Luminous Breakdown in Ruby and Related Effects. ZhETF, 1968, Vol. 54, Issue 1, pp. 37-45.

19. Genkin, V. N., Miller, A. M., Soustov, L. V. The Dynamics of Laser-Induced Damage of KDP Crystals. ZhETF, 1980, Vol. 79, Issue 5, pp. 1880-1887.

20. Gorshkov, B. G., Epifanov, A. S., Manenkov, A. A., Panov, A. A. The Breakdown of Wideband Dielectrics by UV Laser Emission. Kvantovaya elektron, 1979, Vol. 6, Vol. 11, pp. 2415-2419.

21. Ryvkin, S. M. Fotoelektricheskie yavleniya v poluprovodnikakh [The Photoelectric Phenomena in Semiconductors]. Moscow: Fizmatgiz, 1963, 494 pages.

22. Dneprovskiy, V. S., Klyshko, D. N., Penin, A. N. The Photoconductivity of Dielectrics Under Laser Irradiation. Pis'ma v ZhETF, 1966, Vol. 3, Issue 10, pp. 385-389.

23. Aseev, G. I., Kats, M. L., Nikol'skiy, V. K. Multiphoton Excitation of Photoconductivity in Alkali-Halide Crystals Under Laser Irradiation. Pis'ma v ZhETF, 1968, Vol. 8, Issue 4, pp. 174-177.

24. Catalano, I. M., Cingolani, A., Minafra, A. Multiphoton Transitions in Ionic Crystals. Phys. Rev. B - Solid State, 1972, Vol. 5, No. 4, pp. 1629-1632.

25. Williams, R. T., Klein, P. H., Magrwardt, C. L. Multiphoton-Induced Conductivity in Ultraviolet-Transmitting Materials. Laser Induced Damage in Optical Materials, 1977. Wash.: US Government printing office, 1978, pp. 481-488. (NBS spec. publ. 509).

26. Chukova, Yu. P. The Photoconductivity of Diamond in the N 2 Absorption Band. Fizika i tekhnika poluprovodnikov, 1979, Vol. 13, Issue 2, pp. 347-350.

27. Vovilov, V. S., Konorova, E. A., Stepanova, E. B., Trukhan, E. M. The Photoconductivity of Diamond Ion-Doped by Lithium in the IR. Fizika i tekhnika poluprovodnikov, 1979, Vol. 13, Issue 5, pp. 1033-1036.

28. Jain, S. C., Arora, N. D. Photoconductivity of Potassium Colloids in KBr Crystals. J. Phys. and Chem. Solids, 1976, Vol. 37, No. 4, pp. 363-367.

29. Callija, J. M., Agullo-Lopez, F. Photoconductivity of Potassium Colloids in KCl Single Crystals. J. Phys. Soc. Jap., 1974, Vol. 36, No. 3, pp. 739-742.

30. Grishchenko, Yu. I., Pisareva, E. V. The Kinetics of Photoconductivity With Aggregate Color Centers. FTT, 1977, Vol. 10, Issue 4, pp. 1169-1171.

31. Kats, M. L., Gyunsburg, K. E., Golubentseva, L. I., Zvezdova, N. P. The Influence of Oxygen Ions on the Photoconductivity of KCl and KBr Monocrystals. FTT, 1973, Vol. 15, Issue 1, pp. 303-305.

32. Aseev, T. G. Accounting For the Saturation Effect in Multiphoton Ionization Processes of Activated Alkali-Halide Crystals. FTT, 1974, Vol. 16, Issue 1, pp. 293-295.

33. Aseev, G. I., Kach, M. L. Multiphoton Excitation and Ionization of Tl^+ Impurity Centers in Alkali-Halide Crystals. FTT, 1972, Vol. 14, Issue 5, pp. 1365-1368.

34. Aseev, G. I., Kats, M. L., Nikol'skiy, V. K. Multiphoton Ionization of Ag^+ Impurity Centers in a KCl Crystal in a Laser Field. FTT, 1970, Vol. 12, Issue 12, pp. 3393-3399.

35. Aseev, G. I., Kats, M. L., Nikol'skiy, V. K., Elistratov, V. A. Multiphoton Excitation of Luminescence and Photoconductivity of KCl-Eu Monocrystals by Ruby and Neodymium Lasers. Izv. AN SSSR. Ser. fiz., 1969, Vol. 33, No. 5, pp. 858-862.

36. Pologrudov, V. V., Karnaukhov, E. N. The Photoconductivity and Retention of Luminescence Excited in the Longwave Impurity Absorption Bands of Alkali-Halide Crystals. FTT, 1981, Vol. 23, Issue 10, pp. 3033-3037.

37. Bradford, J. N., Williams, R. T., Faust, W. L. Study of F-Center Formation in KCl on a Picosecond Time Scale. Phys. Rev. Lett., 1975, Vol. 35, No. 5, pp. 300-303.

38. Williams, R. T., Bradford, J. N., Faust, W. L. Short-Pulse Optical Studies of Exciton Relaxation in F-Center Formation in NaCl, KCl and NaBr. Phys. Rev. B - Solid State, 1978, Vol. 18, No. 12, pp. 7038-7057.

39. Bredikhin, V. I., Genkin, V. N., Miller, A. M., Soustov, L. V. The Experimental Investigation of the Nature of Photoelectric Phenomena in KDP and DKDP Crystals. ZhETF, 1978, Vol. 75, Issue 5, pp. 1763-1770.

40. Bredikhin, V. I., Genkin, V. N., Miller, A. M., Soustov, L. V. Photoelectric Effects in KDP and DKDP Crystals Under Laser Irradiation. Izv. AN SSSR. Ser. fiz., 1979, Vol. 43, No. 2, pp. 309-312.

41. Gorshkov, B. G., Epifanov, A. S., Manenkov, A. A., Panov, A. A. Experimental Investigations of the Photoconductivity of Wideband Dielectrics Excited by UV Laser Irradiation. ZhETF, 1981, Vol. 81, Issue 4, pp. 1423-1434.

42. Rayzer, Yu. P. Lazernaya iskra i rasprostranenie razryadov [Laser Spark and Charge Propagation]. Moscow: Nauka, 1974, 308 pages.

43. Rayzer, Yu. P. Osnovy sovremennoy fiziki gazorazryadnykh protsessov [The Principles of the Modern Physics of Gas-Discharge Processes]. Moscow: Nauka, 1980, 415 pages.

44. Fersman, I. A., Khazov, L. D. The Damage Mechanism to a Transparent Dielectric Surface From Irradiation by a Narrow Light Pulse. Kvantovaya elektron, 1972, No. 4, pp. 25-31.

45. Ohmori, Y., Yasojima, Y., Imuishi, Y. Photoconduction, Thermally Stimulated Luminescence and Optical Damage in Single Crystal of $LiNbO_3$. J. Appl. Phys. Jap., 1975, Vol. 14, No. 9, pp. 1291-1300.

46. Pashkov, V. A., Solov'eva, N. M., Uyukin, E. M. Photo- and Thermal-Conductivities in Lithium Niobate Crystals. FTT, 1979, Vol. 21, Issue 6, pp. 1879-1882.

47. Kabler, M. N., Williams, R. T. Vacancy-Interstitual Pair Production Via Electron-Hole Recombination in Halide Crystals. Phys. Rev. B - Solid State, 1978, Vol. 18, No. 4, pp. 1948-1960.

48. Gutman, F., Layons, L. Organicheskie poluprovodniki [Organic Semiconductors]. Moscow: Mir, 1970, 696 pages.

49. Liu, P., Smith, W., Lotem, H. et al. Absolute Two-Photon Absorption Coefficients at 355 and 266 nm. Phys. Rev. B - Solid State, 1978, Vol. 17, No. 2, pp. 4620-4632.

50. Smith, W. L. Laser Induced Breakdown in Optical Materials. Opt. Eng., 1978, Vol. 17, No. 5, pp. 489-503.

51. Keldysh, L. V. Ionization in a Strong Electromagnetic Wave Field. ZhETF, 1964, Vol. 47, Issue 5, pp. 1945-1957.

52. Dorozhkin, L. M., Doroshenko, V. S., Lazarev, V. V. et al. Recording Pulsed Radiation by Thin-Film Pyroelectric Detectors. Impul'snaya fotometriya. Leningrad: Mashinostroenie, 1979, Issue 6, pp. 64-67.

53. Danileyko, Yu. K., Manenkov, A. A., Prokhorov, A. M., Khaimov-Mal'kov, V. Ya. Surface Damage to Crystals by Ruby Laser Irradiation. ZhETF, 1970, Vol. 58, Issue 1, pp. 31-36.

54. Kepler, R. G. Charge Carrier Production and Mobility in Anthracene Crystals. Phys. Rev.. 1960, Vol. 119, No. 4, pp. 1226-1229.

55. Gorshkov, B. G., Epiphanov, A. S., Manenkov, A. A., Panov, A. A. Studies of Laser-Produced Damage to Transparent Optical Materials in the UV Region and in Crossed UV-IR Beams. Laser Induced Damage in Optical Materials, 1981. Wash.: US Government printing office, 1983, pp. 76-86. (NBC spec. publ. 638).

56. Mari, Dzh., Donzhon, Zh. Monocrystal Ferroelectric and Their Application in Light-Activated Data Display Devices. TIIER, 1973, Vol. 61, No. 7, pp. 178-197.

57. Gorshkov, B. G. Issledovanie mekhanizmov razrusheniya ionnykh kristallov pod deystviem impul'snogo lazernogo izlucheniya nanosekundnogo diapazona [An Investigation of the Breakdown Mechanisms of Ionic Crystals Under Nanosecond Pulsed Laser Irradiation]. Metod MBR v kvantovoj elektronike i lazernoe razrushenie [The Brillouin Method in Quantum Electronics and Laser-Induced Damage]. Moscow: Nauka, 1982, pp. 81-134. (Tr. FIAN; Vol. 137).

58. Voronkova, E. M., Grechushnikov, B. N., Distler, G. I., Petrov, I. P. Opticheskie materialy dlya infrakrasnoj tekhniki [Optical Materials For Infrared Engineering]. Moscow: Nauka, 1965, 335 pages.

59. Miuta, T., Tomiki, T. Optical Studies of NaCl Single Crystals in 10 eV Region. II J. Phys. Soc. Jap., 1968, Vol. 24, No. 6, pp. 1286-1301.

60. Aluker, E. D., Lusis, D. Yu., Chernov, S. A. Elektronnye vozbuzhdeniya i radiolyuminestsentsiya shchelochnogaloidnykh kristallov [Electron States and the Radio Luminescence of Alkali-Halide Crystals]. Riga: Einatne, 1979, 251 pages.

61. Piacentini, M., Lynch, D. M., Olson, C. J. Thermoreflectance of LiF Between 12 and 30 eV. Phys. Rev. B - Solid State, 1976, Vol. 13, No. 12, pp. 5530-5543.

62. Tomiki, T., Miyata, T., Tsukamoto, H. Temperature Dependence of the Fundamental Spectra of Potassium Halides in the Schuman Ultraviolet Region (4.4-13.5 eV). J. Phys. Soc. Jap., 1973, Vol. 35, No. 2, pp. 495-507.

63. Hopfield, J. J., Worlock, J. M. Two Quantum Absorption Spectra of KI and CsI. Phys. Rev. A - Gen. Phys., 1965, Vol. 137, No. 5, pp. 1455-1464.

64. Froehlich, D., Stuginnuss, B. New Assignments of the Bandgap in Alkali Bromides by Two-Photon Spectroscopy. Phys. Rev. Lett., 1967, Vol. 19, No. 9, pp. 496-498.

65. Froehlich, D., Stuginnuss, B., Onadera, Y. Two-Photon Spectroscopy in CsI and CsBr. Phys. status solidi., 1970, Vol. 40, No. 2, pp. 547-556.

66. Pong, W., Smith, J. A. Photoemission Studies of LiCl, NaCl and KCl. Phys. Rev. B - Solid State, 1974, Vol. 9, No. 6, pp. 2674-2677.

67. Poole, R. T., Jenkin, J. G., Liesegang, J., Leckey, R. C. G. Electronic Band Structure of the Alkali-Halides. I. Experimental Parameters. Phys. Rev. B - Solid State, 1975, Vol. 11, No. 12, pp. 5179-5189.

68. Poole, R. T., Liesegang, J., Leckey, R. C. G., Jenkin, J. G. Electronic Band Structure of the Alkali-Halides. II. Critical Survey of the Theoretical Calculations. Phys. Rev. B - Solid State, 1975, Vol. 11, No. 12, pp. 5190-5196.

69. Volkova, N. V. The Influence of Shortwave Absorption on the Breakdown Threshold of Optical Crystals Under Light Irradiation. FTT, 1974, Vol. 16, Issue 1, pp. 307-308.

70. Blechsmidt, D., Skibowski, M., Steimann, W. Photoemission From Potassium Halides in the Photon Energy Range 7 to 30 eV. Phys. status solidi (B), 1970, Vol. 42, No. 1, pp. 61-70.

71. Roesler, D. M., Walker, W. C. Electronic Spectra of Crystalline NaCl and KCl. Phys. Rev., 1968, Vol. 166, No. 3, pp. 599-606.

72. Baldini, G., Bossacchi, B. Optical Properties of Alkali-Halide Crystals. Phys. Rev., 1968, Vol. 166, No. 3, pp. 863-870.

73. Teegarden, K., Baldini, G. Optical Absorption Spectra of Alkali Halides at 10° K. Phys. Rev., 1967, Vol. 155, No. 3, pp. 896-907.

74. Kunz, A. B. Electronic Bands For Rubidium Chloride and Face Centered Cubic Alkali Bromides. Phys. Status Solidi, 1968, Vol. 29, No. 1, pp. 115-120.

75. Nikogosyan, D. N. Crystals For Nonlinear Optics: (Reference Survey). Kvantovaya elektron, 1977, Vol. 4, No. 1, pp. 5-26.

76. Gulamov, A. A., Ibragimov, E. A., Redkoretsev, V. I., Usmanov, T. The Maximum Lasing Efficiency of the Second and Third Harmonics From a Neodymium Laser. Kvantovaya elektron, 1983, Vol. 10, No. 7, pp. 1305-1306.

77. Saito, S., Onaka, R. Electronic Structures of KDP and Its Family. Ferroelectrics, 1978, Vol. 21, No. 2, pp. 553-554.

78. Ahrenkiel, R. K., Brown, F. C. Electron Hall Mobility in the Alkali Halides. Phys. Rev. A - Gen. Phys., 1964, Vol. 136, No. 1, pp. 223-231.

79. Bass, M., Franken, P. A., Ward, J. F. Optical Rectification. Phys. Rev. A - Gen. Phys., 1965, Vol. 138, No. 2, pp. 534-542.

80. Hochuli, U. E. Photoconductivity Measurements in Ruby and Sapphire. Phys. Rev. A - Gen. Phys., 1964, Vol. 133, No. 2, pp. 468-471.

81. Kikoin, I. K. Tablitsy fizicheskikh velichin [Tables of Physical Quantities]. Moscow: Atomizdat, 1976, 1006 pages.

82. Gurevich, L. E., Rumyantsev, A. A. The Theory of the Electrooptic Effect in Organic Crystals at High Frequencies and in an External Magnetic Field. FTT, 1967, Vol. 9, Issue 1, pp. 75-78.

83. Valov, P. M., Danishevskiy, A. M., Kastal'skiy, A. A. et al. Photon Drag on Electrons With Intraband Light Absorption by Free Current Carriers in Semiconductors. ZhETF, 1970, Vol. 59, Issue 6, pp. 1919-1925.

84. Danishevskiy, A. M., Kastal'skiy, A. A., Ryvkin, S. M., Yaroshetskiy, I. D. Photon Drag on Three Carriers in Direct Band-to-Band Transitions in Semiconductors. ZhETF, 1970, Vol. 58, Issue 2, pp. 544-550.

85. Valov, P. M., Ryvkin, B. S., Ryvkin, S. M. et al. Light Drag on Electrons in the Photoionization of Impurity Centers. Fizika i tekhnika poluprovodnikov, 1971, Vol. 5, Issue 9, pp. 1772-1775.

86. Gorshkov, B. G., Panov, A. A. Photon Drag on Electrons in Alkali-Halide Crystals. FTT, 1981, Vol. 23, Issue 12, pp. 3597-3601.

87. Smith, W. L., Bechtel, J. H., Bloembergen, N. Dielectric Breakdown Threshold and Nonlinear Refractive Index Measurements With Picosecond Laser Pulses. Phys. Rev. B - Solid State, 1975, Vol. 12, No. 2, pp. 706-717.

88. Scinner, D. K., Witcher, R. E. Measurements of the Radius of a High Power Laser Beam Near the Focus of a Lens. J. Phys. E: Sci. Instrum., 1972, Vol. 5, No. 1, pp. 237-238.

89. Grinberg, A. A., Makovskiy, L. L. The Theory of Photoelectric and Photomagnetic Effects Caused by the Photon Momentum in the Photoionization of Impurity Centers in Semiconductors. Fizika i tekhnika poluprovodnikov, 1970, Vol. 4, Issue 6, pp. 1162-1167.

90. Harper, P. G., Hodby, J. W., Stradling, R. A. Electrons and Optic Phonons Electronic Excitation in Solids - The Effects of Longitudinal Optical Lattice Vibrations on the Electronic Excitations of Solids. Rep. Progr. Phys., 1973, Vol. 36, No. 1, pp. 1-102.

91. Berkovskiy, F. I., Ryvkin, S. M. The Sensitivity of Germanium and Silicon Elements in the Impurity Excitation Range. FTT, 1962, Vol. 4, Issue 2, pp. 366-375.

92. Bonch-Bruevich, V. L., Kalashnikov, S. G. Fizika poluprovodnikov [The Physics of Semiconductors]. Moscow: Nauka, 1977, 672 pages.

93. Epifanov, A. S., Manenkov, A. A., Panov, A. A., Shakhverdiev, E. M. A Kinetic Analysis of UV Laser Photoconductivity in Alkali-Halide Crystals. Tez. dokl. VI Vsesoyuz. konf. po nerezonansnomu vzaimodey-

stviyu opticheskogo izlucheniya s veshchestvom, Palanga, 19-21 sent. [Topic Papers of the 7th All-Union Conference on Nonresonant Inter action of Optical Emission With Matter, Palanga, 19-21 September]. 1984, Vil'nyus, 1984, pp. 190-191.

94. Lyamov, V. E., Madvaliev, U., Shikhlinskaya, R. E. The Photoacoustic Spectroscopy of Solids. Akust. zhurn., 1979, Vol. 25, No. 3, pp. 427-433.

95. Scryby, C. B., Collingwood, J. C., Wadley, H. N. G. A New Technique For the Measurements of Acoustic Emission Transients and Their Relationship to Crack Propagation. J. Phys. D: Appl. Phys., 1978, Vol. 11, No. 17, pp. 2359-2369.

96. Hordvic, A., Schlossberg, H. Photoacoustic Technique For Determining Optical Absorptions Coefficients in Solids. Appl. Opt., 1977, Vol. 16, No. 1, pp. 101-107.

97. Rosencwaig, A., Gersho, A. Theory of Photoacoustic Effect With Solids. J. Appl. Phys., 1976, Vol. 47, No. 1, pp. 64-69.

98. Inagaki, T., Kagami, K., Arakawa, E. T. Photoacoustic Study of Surface Plasmons in Metals. Appl. Opt., 1982, Vol. 21, No. 5, pp. 949-954.

99. Gorshkov, B. G., Dorozhkin, L. M., Epifanov, A. S. et al. An Investigation of Two-Photon Light Absorption in Alkali-Halide Crystals by the Optoacoustic Technique. Tez. dokl. VI Vsesoyuz. konf. po nerezonansnomu vzaimodeystviyu opticheskogo izlucheniya s veshchestvom, Palanga, 19-21 sent. [Topic Papers of the 6th All-Union Conference on the Nonresonant Interaction of Optical Emission With Matter, Palanga, 19-21 September]. 1984, Vil'nyus, 1984, pp. 149-150.

100. Gorshkov, B. G., Dorozhkin, L. M., Epifanov, A. S. et al. An Investigation of Nonlinear UV Laser Emission Absorption in Alkali-Halide Crystals in Generating Nonequilibrium Carriers by the Opto-acoustic Techniques. ZhETF, 1985, Vol. 88, Issue 1, pp. 21-29.

101. Voronov, F. F., Chernysheva, E. V., Goncharova, V. A. The Elastic Properties of an NaCl Monocrystal at Pressures up to 9 GPa and a Temperature of 293 K. FTT, 1979, Vol. 21, Issue 1, pp. 100-105.

102. Dzhavodov, L. N., Krotov, Yu. I. The Influence of Pressure on the Thermal Conductivity of NaCl, KCl and RbCl. FTT, 1978, Vol. 20, Issue 3, pp. 654-657.

SUBJECT INDEX